DEWALT®

ELECTRICAL
CODE REFERENCE
Second Edition

Written by
Daniel Sandefur
and
American Contractor's Exam Services

Published by

DELMAR
CENGAGE Learning™

dewalt.com/guides

DELMAR
CENGAGE Learning™

DeWALT Electrical Code Reference Second Edition: Based on the National Electrical Code® including the NEC® 2011

Daniel Sandefur and American Contractors Exam Services

Vice President, Technology and Trade Professional Business Unit: Gregory L. Clayton

Director of Building Trades: Taryn Zlatin McKenzie

Product Development Manager: Robert Person

Development: Nobina Chakraborti

Director of Marketing: Beth A. Lutz

Marketing Manager: Marissa Maiella

Production Director: Carolyn Miller

Production Manager: Andrew Crouth

Art Director: Benjamin Gleeksman

Content Project Management: PreMediaGlobal

Compositor: PreMediaGlobal

For product information and technology assistance, contact us at **Cengage Learning Customer & Sales Support, 1-800-354-9706**

For permission to use material from this text or product, submit all requests online at **www.cengage.com/permissions**
Further permissions questions can be e-mailed to **permissionrequest@cengage.com**

Library of Congress Control Number: 2010939880

ISBN-13: 978-1-111-54548-2
ISBN-10: 1-111-54548-0

Delmar
5 Maxwell Drive, Clifton Park, NY 12065-2919, USA

Cengage Learning is a leading provider of customized learning solutions with office locations around the globe, including Singapore, the United Kingdom, Australia, Mexico, Brazil, and Japan. Locate your local office at: **international.cengage.com/region**

Cengage Learning products are represented in Canada by Nelson Education, Ltd.

Visit us at www.InformationDestination.com
For more learning solutions, please visit our corporate website at **www.cengage.com**

Printed in China
3 4 5 6 7 14 13

A Note to the Reader

The DEWALT Electrical Code Reference is not a formal code interpretation and is not intended to replace the code. You will need a copy of the 2008 and 2011 NFPA 70™ National Electrical Code® (NEC®) to check all references given in this book.

This book is designed to help you understand some, but not all, phases of electrical installations. The guidelines illustrated in this book are not the only way to install the systems shown. Local codes sometimes have exceptions or regulations that are enforced along with the 2008 and 2011 NEC®.

Anyone working on an electrical project should contact their local building inspection department, office of planning and zoning, and/or department of permits in order to learn which codes are being used and how they will affect their project.

Both the 2008 and 2011 NEC Code are copyrighted work owned by the National Fire Protection Association®.

"NFPA 70" is a trademark of the National Fire Protection Association.

"NFPA, National Fire Protection Association, National Electrical Code, and NEC" are registered trademarks of the National Fire Protection Association.

For more information on the National Fire Protection Association or the National Electrical Code, please contact:

Phone: 1-800-344-3555

Web: www.nfpa.org

CONTENTS

AFCI (Arc-Fault Circuit Interrupter)

A device intended to provide protection from the effects of arc faults by recognizing characteristics unique to arcing and by functioning to de-energize the circuit when an arc fault is detected. **Example:** Arc faults may occur when loose connections or faulty appliances are introduced into a circuit. This type of circuit protection is intended to prevent fires caused by these types of faults.

Branch Circuit

The circuit conductors between the final overcurrent device protecting the circuit and the outlet(s).

Current

A measurement of the rate of flow of electricity through a conductor, measured in amps.

Device

A part of an electrical system that is intended to carry or control but not utilize electricity (*i.e., switches, thermostats*).

Equipment Grounding Conductor

A conductor used to ground electrical equipment. Commonly referred to in the electrical field as the "ground." These conductors are usually identified by green insulation or bare. The NEC uses the term "equipment grounding conductor" for these conductors.

Feeder

The circuit conductors between the service and the final overcurrent devices. Subpanels are commonly located throughout buildings and the circuits feeding these panels are called feeders. Simply put, the conductors between 2 overcurrent devices are called feeder conductors.

GFCI (Ground-Fault Circuit Interrupter)

A device intended for the protection of personnel that functions to de-energize a circuit or portion of a circuit when the leakage current falls within a range of 4–6 milliamps. **Example:** Ground-fault circuit interrupters are used to provide protection to persons. These devices are set to trip at a threshold below what would be harmful to people. These devices are usually required where water and electricity may both be present.

Grounded Conductor

A conductor which is grounded. Commonly referred to in the electrical field as the "neutral" conductor. These conductors are usually identified by white or grey insulation. The NEC uses the term "grounded conductor" for these conductors.

Grounding Electrode

A device that establishes an electrical connection to the earth. Examples of a grounding electrode include rebar in the footer, metal underground water piping with 10' or more in contact with the earth, and ground rods.

Grounding Electrode Conductor

A conductor used to connect the grounding electrode to the equipment grounding conductor, grounded conductor or both at the service, at each building or structure where supplied by a feeder or branch circuit, or at the source of a separately derived system. These are the conductors that are used for grounding at the service or separate building. They are used to connect to the grounding electrode system which includes metal water piping, rebar in footers, and ground rods.

Line

The designation "line" used in electrical terms refers to the incoming power. **Example:** Most panelboards and disconnects will be labeled "line" and "load." The line side of the equipment will be where the source of power is terminated.

Listed

Evaluated and tested by third-party agency such as UL, ETL, MET Labs, etc.

Load

The designation "load" used in electrical terms refers to the outgoing power. ***Example:*** Most panelboards and disconnects will be labeled "line" and "load." The load side of the equipment will be where the outgoing power is terminated.

Outlet

A point on the wiring system where current is taken to supply equipment. ***Example:*** Receptacles, lighting fixtures, smoke detectors and appliances are included in this definition, but a switch would not since no current is taken at a switch, it simply passes through.

Overcurrent Protective Devices

A device set to open a circuit when the current exceeds a set value. Usually consists of circuit breakers or fuses.

Resistance

A measurement of the opposition to the flow of electricity through a conductor, measured in ohms.

Romex

Trade name for type NM cable. This is the most commonly used cable in residential wiring. The proper name for this cable is nonmetallic sheathed cable or NM cable. This cable comes in a variety of sizes and number of conductors. ***Example:*** 12/2 with ground would contain 3 – 12 gauge conductors, 1 black insulated conductor, 1 white insulated conductor and 1 bare conductor. 12/3 with ground would contain 4 – 12 gauge conductors, 1 black insulated conductor, 1 red insulated conductor, 1 white insulated conductor and 1 bare conductor.

Service

The conductors and equipment for delivering electricity from the utility company to the wiring system of the premises served. Usually referred to by electricians as the "main" panel. The service is the first point of disconnect for the conductors from the utility company.

SE Cable

Service entrance cable is not commonly used for service cable any longer, but is used for branch circuit and feeder wiring in residential applications. Much like NM cable SE cable comes in a wide variety of wire sizes. SEU cable usually contains 3 conductors, 2 which will be insulated and a bare equipment grounding conductor. Type SER cable usually contains 4 conductors, 3 that will be insulated and 1 bare equipment grounding conductor. ***Example:*** #2/2/2/4 aluminum SER cable will contain 2 black insulated #2 aluminum conductors (some have 1 black and 1 red), 1 white insulated #2 aluminum conductor and 1 bare #4 equipment grounding conductor. #2/2/4 aluminum SEU cable will contain 2 black insulated #2 aluminum conductors (some have 1 black and 1 red) and 1–#4 bare equipment grounding conductor.

Ungrounded Conductor

A conductor that is not grounded. Commonly referred to in the electrical field as the "hot" or "line" conductor. These conductors are usually identified by black or blue (any color other than green, white, or gray) insulation. The NEC uses the term "ungrounded conductor" for these conductors.

Voltage

A measurement of the force at which the electrons are moved through a conductor.

Wattage

A measurement of the rate of consumption of electricity.

BASIC ELECTRICAL EQUIPMENT

Application	Description	Equipment
	CABLE CONNECTORS	
	312.5(C) & 314.17(B) **Plastic Push-in Cable Connectors.** Used to secure NM cables to metallic boxes and panelboards. Connector simply pushes into knockout and grips to secure the cable.	
	312.5(C) & 314.17(B) **Squeeze Type Cable Connector.** Used to secure NM cables to metallic boxes and panelboards. Connector is inserted into knockout and squeezed to elongate and secure to cabinet.	
	312.5(C) & 314.17(B) **Locknut Type Cable Connector.** Used to secure NM cables to metallic boxes and panelboards. Connector is inserted into knock-out and locknut must be installed on the opposite side. Screw type clamping mechanism secures cable.	
	FILLER PLATES	
	110.12(A) **Knockout Filler.** Used to fill in unused holes in electrical equipment. Insert into hole and bend tabs over on opposite side to prevent removal.	
	408.7 **Circuit Breaker Knockout Filler.** Used to fill in unused holes in panelboard covers. Insert into hole and tabs prevent removal.	

> **YOU SHOULD KNOW:**

- Most cable connectors are only listed for use with one large cable (#10 or larger) or two smaller cables (#14–#12).

BASIC ELECTRICAL EQUIPMENT (*cont.*)

Application	Description	Equipment
	WIRE CONNECTORS	

110.14(B)

Wire Connectors. Commonly referred to as wire nuts, these devices are used to splice multiple conductors together. Simply twist conductors together and tighten wire connector over wires. Come in a wide range of sizes indicated by multiple colors. Generally orange connectors are good for 2 small conductors, yellow connectors are good for 3 small conductors and red connectors are good for up to 5 conductors. Tan connectors can accommodate a wider range of conductor sizes.

250.8 & 250.148

Grounding Wire Connectors. Used to splice multiple grounding conductors while leaving provisions for grounding devices such as receptacles or switches. Twist grounding conductors together, cut off unused grounding conductor and then install wire connector or splice cap.

| | **CABLE STAPLES** | |

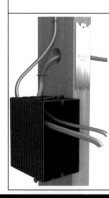

334.30 & 314.17(C)

Cable Staples. Used to secure cables routed across or through framing members. May be used on many different cable types including NM, coaxial, or telephone wiring. Staple is driven into wood framing members with a hammer carefully to avoid damaging the cable.

Staples come in a wide variety of sizes to accommodate different cable sizes.

PERMISSIBLE RECEPTACLE TYPES
FOR BRANCH CIRCUITS – TABLE 210.21(B)

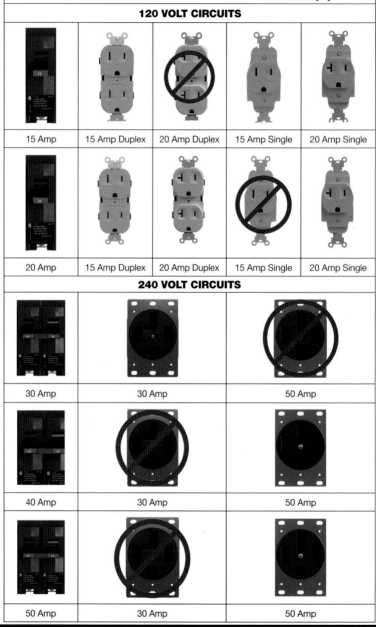

120 VOLT CIRCUITS

15 Amp	15 Amp Duplex	20 Amp Duplex	15 Amp Single	20 Amp Single
20 Amp	15 Amp Duplex	20 Amp Duplex	15 Amp Single	20 Amp Single

240 VOLT CIRCUITS

30 Amp	30 Amp	50 Amp
40 Amp	30 Amp	50 Amp
50 Amp	30 Amp	50 Amp

TYPES OF TERMINALS

SCREW TERMINALS — 110.14(A)

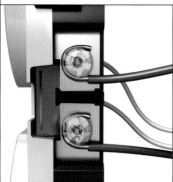

Feature wire binding screws where the conductors are wrapped around the screw. The conductors should be wrapped clockwise and make full contact with the screw. This connection type is limited to #10 or smaller conductors.

PUSH-IN TERMINALS

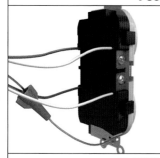

Feature push-in back-wire connections that are made by stripping approximately 3/4" of insulation from the conductor and inserting it into the hole provided on the rear of the receptacle. This connection is typically provided for #14 conductors only.

CLAMPING TYPE TERMINALS

Feature a pressure plate behind the screw terminal. Connection is made by stripping approximately 1" of insulation from the conductor and inserting into the hole provided in the rear of the receptacle and tightening the screw. The plate is tightened against the conductor to make the connection.

▶ **YOU SHOULD KNOW: BE CAREFUL**
- Do NOT remove too much insulation.
- Check for proper connections.

1 POLE SWITCH

For illustration clarification, grounding conductors are not shown.

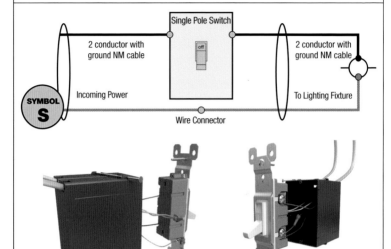

No.	Description
1	Supply power to the switch with 2 conductor with ground NM cable from the power source *(panelboard, receptacles, etc.)*.
2	Supply power to the light with 2 conductor with ground NM cable from the switch outlet box.
3	Join neutrals (white wires) in the switch box with wire connectors. *(Do Not switch neutrals, 404.2(B).)*
4	Connect black wires to gold screws on switch.
5	Join grounds and connect to green grounding screw on switch.
6	The switch must be oriented so that the "on" position is up and the "off" position is down. *(404.7)*
7	Connect lighting fixture to wires from 2 conductor with ground NM cable at the lighting fixture outlet box.

3 WAY SWITCH

For illustration clarification, grounding conductors are not shown.

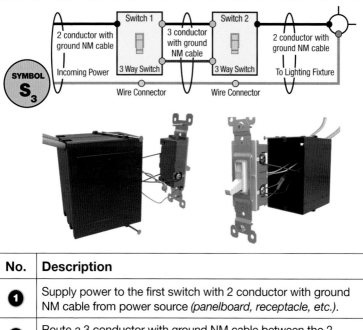

No.	Description
1	Supply power to the first switch with 2 conductor with ground NM cable from power source *(panelboard, receptacle, etc.)*.
2	Route a 3 conductor with ground NM cable between the 2 switch outlet boxes.
3	Supply power to the light with 2 conductor with ground NM cable from the second switch outlet box.
4	Join neutral (white) wires in both switch boxes with wire connectors. *(Do Not switch neutrals 404.2(B).)*
5	Connect the black wire from the 2 conductor with ground NM cable to the common (black) screw on each 3 way switch.
6	Connect the black and red wires from each 3 conductor with ground to the gold screws on each 3 way switch. *(Either wire may be connected to either screw.)*
7	Join grounds in each box and connect to green grounding screw on each switch.
8	Connect lighting fixture to wires from 2 conductor with ground NM cable at the lighting fixture outlet box.

4 WAY SWITCH

For illustration clarification, grounding conductors are not shown.

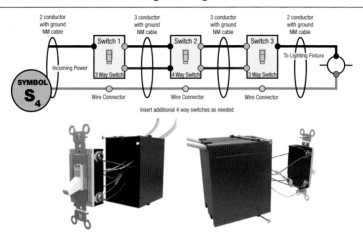

No.	Description
1	Supply power to the first switch with 2 conductor with ground NM cable from power source *(panelboard, receptacle, etc.)*.
2	Route a 3 conductor with ground NM cable from 1st switch box to each successive switch box.
3	Supply power to the light with 2 conductor with ground NM cable from the last switch outlet box.
4	Join neutral (white) wires in all switch boxes with wire connectors. *(Do Not switch neutrals 404.2(B).)*
5	In the first and last switch boxes, see wiring diagram for a 3 way switch on page 8.
6	In the middle switch boxes connect the red and black wires from one 3 conductor with ground NM cable to the terminals marked input, and the red and black wires from the other 3 conductor with ground NM cable to the terminals marked output.
7	Join grounds in each box and connect to green grounding screw on each switch.
8	Connect lighting fixture to wires from 2 conductor with ground NM cable at the lighting fixture outlet box.

120 VOLT RECEPTACLE

For illustration clarification, grounding conductors are not shown.

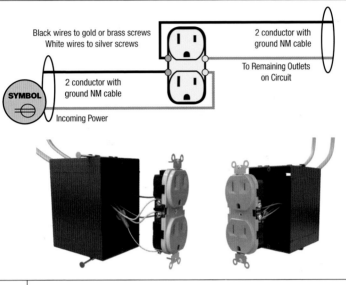

Black wires to gold or brass screws
White wires to silver screws

2 conductor with
ground NM cable

To Remaining Outlets
on Circuit

SYMBOL

2 conductor with
ground NM cable

Incoming Power

No.	Description
❶	Used throughout the home to supply power to appliances.
❷	Generally powered with 2 conductor with ground NM cable.
❸	If multiple receptacles are installed on the same circuit simply feed through the first receptacle to the remaining receptacles on the circuit until the circuit is complete.
❹	Black wires (hots) terminate on the gold screws, they do not need to be joined with wire connectors, the receptacle will be used to connect the wires.
❺	White wires (neutrals) terminate on the silver screws they do not need to be joined with wire connectors, the receptacle will be used to connect the wires.
❻	Bare grounding wires should be joined and terminate on the green grounding screws.

120 VOLT GFCI RECEPTACLE

For illustration clarification, grounding conductors are not shown.

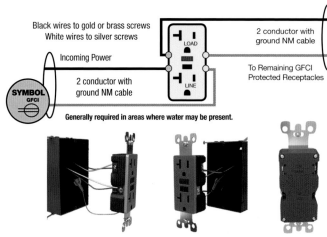

Black wires to gold or brass screws
White wires to silver screws

Incoming Power

2 conductor with
ground NM cable

LOAD

LINE

2 conductor with
ground NM cable

To Remaining GFCI
Protected Receptacles

SYMBOL
GFCI

Generally required in areas where water may be present.

No.	Description
①	Generally powered with 2 conductor with ground NM cable.
②	If multiple receptacles are installed on the same circuit and require GFCI protection simply feed through the first receptacle to the remaining receptacles on the circuit until the circuit is complete. (*GFCI receptacles are not required on the remaining receptacles, protection will be provided by the first receptacle.*)
③	Black wire (hot) from power supply will be connected to the gold screw marked "Line".
④	White wire (neutral) from power supply will be connected to the silver screw marked "Line".
⑤	If multiple receptacles are installed on the circuit, the black wire to the remaining receptacles will connect to the gold screw marked "Load".
⑥	If multiple receptacles are installed on the circuit, the white wire to the remaining receptacles will connect to the silver screw marked "Load".
⑦	Bare grounding wires should be joined and terminate on the green grounding screw.

4 WIRE 240 VOLT RECEPTACLE

Grounding conductor may be green or bare.

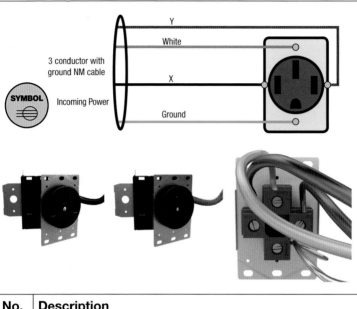

3 conductor with ground NM cable

SYMBOL

Incoming Power

Y
White
X
Ground

No.	Description
❶	Generally used for larger appliances such as ranges and clothes dryers.
❷	Only 1 receptacle is usually installed on each circuit.
❸	Generally supplied with a 3 conductor with ground type NM cable or 4 conductor type SER cable.
❹	Terminals may not be color coded such as with 120 volt receptacles, but will be marked with some designation.
❺	Red and black wires (hots) terminate on screws marked "Line", "X" , "Y" or a similar marking. (*Either wire may terminate on either terminal.*)
❻	White wires (neutrals) terminate on the terminal marked "Neutral", "N", "WH", "White" or a similar marking.
❼	Green or bare grounding wires terminate on the terminal marked "G", "Gr", or "Ground".

LIGHTING FIXTURE

There are two different methods for wiring a lighting fixture.

Power Fed to the Switch	Power Fed to the Light

No.	Description
1	From panelboard or receptacle (power supply) with 2 conductor with ground NM cable *(must be supplied from a maximum 120 volt circuit)*. **210.6(A)**
2	From switch outlet box to lighting outlet box *(2 conductor with ground NM cable)*.
3	Connect black power supply wire to white wire from switch outlet box.
4	White wire must be identified as a "Hot" *(typically identified with black electrical tape at each termination point)*. **200.7(C)(2)**
5	Connect matching colors *(white to white, black to black, ground to ground)*.
6	Connect insulated wires to switch terminals and bare conductor to grounding terminal.
7	Connect the two white wires together and join with wire connector. Connect two black wires to switch terminals. Join two grounding conductors and ground switch.

CEILING FAN WITH LIGHT

Light and Fan (switched separately)

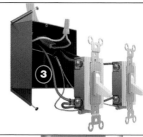
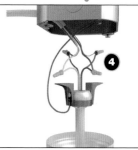
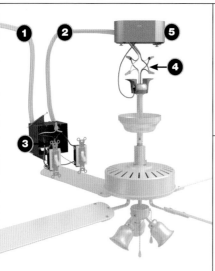

No.	Description
1	From panelboard or receptacle (power supply) with 2 conductor with ground NM cable *(must be supplied from a maximum 120 volt circuit)*. **210.6(A)**
2	From switch outlet box to lighting outlet box *(3 conductor with ground NM cable)*.
3	Connect the two white wires together and join with wire connector. Connect black power supply wire to both switches. Connect black wire to fan to one switch and red wire to light on the fan to other switch. Join grounding conductors and ground both switches.
4	Connect grounding conductors and ground outlet box if metallic. Connect white wires together, black wires together, connect red wire from switch box to the light kit *(commonly blue wire but will be identified with a tag)*.
5	Must use outlet box listed for support of ceiling fans. **314.27(D)**

COMMON CABLE TYPES FOR VARIOUS CIRCUITS

	15 Amp	20 Amp	30 Amp	40 Amp	50 Amp
Plugs & Switches: Living Area	#14/2 with ground Copper N.M	#12/2 with ground Copper N.M	X	X	X
Plugs & Switches: Kitchen, Laundry, Bathrooms, and Dining Rooms	X	#12/2 with ground Copper N.M	X	X	X
Microwave	X	#12/2 with ground Copper N.M	X	X	X
Dishwasher/ Disposal	#14/2 with ground Copper N.M	#12/2 with ground Copper N.M	X	X	X
Refrigerator/ Freezer	#14/2 with ground Copper N.M	#12/2 with ground Copper N.M	X	X	X
Oven/ Cooktop	X	X	#10/3 with ground Copper N.M	#8/3 with ground Copper N.M / 6/6/6/6 AL SER	#6/3 with ground Copper N.M / 4/4/4/6 AL SER
Free Standing Range	X	X	X	#8/3 with ground Copper N.M / 6/6/6/6 AL SER	#6/3 with ground Copper N.M / 4/4/4/6 AL SER
Water Heater	X	X	#10/2 with ground Copper N.M	X	X
Clothes Dryer	X	X	#10/3 with ground Copper N.M	X	X
Hydromassage Tub	X	#12/2 with ground Copper N.M	X	X	X

SIZING CONDUCTORS AND OVERCURRENT PROTECTION

Step	Directions
1	Determine the wattage and voltage of the appliance. Information will be found on appliance nameplate. *(1 kw = 1,000 watts, for example: 4.5 kw = 4,500 watts)*
2	Determine the current of the appliance by dividing the wattage by the voltage. *(4,500W divided by 240V = 18.75 amps)*
3	Determine the MAXIMUM overcurrent protection and conductor size for the appliance by multiplying the current by 150%. *(422.11(E) which applies to a single non-motor operated appliance such as water heaters and ovens.) (18.75 multiplied by 150% = 28.125 amps)*
4	If the calculation results in a "non-standard Overcurrent Device" such is the case in this example, you may go up to, but not exceed, the next standard overcurrent device. *(Refer to table below for proper circuit breaker and wire sizes.)*

Step 1 nameplate:

ELECTRIC STORAGE TANK WATER HEATER

MODEL NUMBER	SERIAL NUMBER	ITEM ID/PART NUMBER
GCV 40 100	AC040034553	9211681000

VOLTS	TOTAL CONNECTED WATTS	WATTS LOWER	CAPACITY U.S. GAL.	MAX WORKING PRESSURE
240	4500	4500	50.0	150

VOLTS	TOTAL CONNECTED WATTS
240	4,500

	Circuit Size (in amps)	Copper NM Cable	Copper SE Cable	Aluminum SE Cable
Standard Overcurrent Device Sizes	15	#14	—	—
	20	#12	—	—
	25–30	#10	—	—
	35–40	#8	#8	#6
	45–60	#6	#6	#4
	70	#4	#4	#3
	80	#3	#3	#2
	90	#3	#3	#1
	100	#2	#2	1/0
	110	—	#1	2/0
	125	—	1/0	3/0
	150	—	2/0	4/0
	175	—	3/0	250 kcmil
	200	—	4/0	300 kcmil

Table is based on excerpts from 240.4(B), 240.4(D), 240.6, Table 310.16, 334.80, and 338.10(B)(4)(a).

CABLES AND OUTLET BOXES INSTALLATION

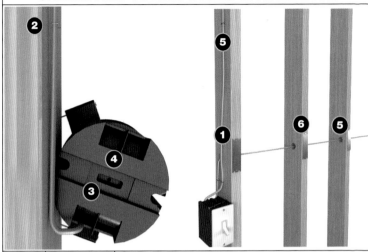

No.	Code	Description
1	334.30	Flat cables must be stapled on the flat side. Do NOT staple flat cables on the edge.
2	334.30	Cable must be secured within 12" of the outlet box.
3	314.17(C)	Cables must be secured to the outlet box. Do NOT remove built-in cable clamps.
4	314.27(A)	Outlet boxes must be listed for support of lighting fixtures. *(Typically round boxes.)*
5	334.30 & 334.30(A)	Cables must be secured every 4' 6". Cables routed horizontally through bored holes in framing members do NOT require additional support.
6	334.80 & 338.10(B) (4)(a)	When 3 or more NM or SE cables are installed in contact with insulation without maintaining spacing or pass through the same opening in wood framing members that are to be caulked or sealed, the allowable ampacity of each conductor must be adjusted in accordance with NEC Table 310.15(B)(2)(A).

> **YOU SHOULD KNOW: NEC 220.14(J)**

- Code does not mandate a maximum number of receptacle outlets on residential circuits. However, typically 8–10 devices *(lights, receptacles and smoke detectors)* are installed on each branch circuit. Appliances and higher wattage lighting fixtures *(such as chandeliers)* should be installed on individual circuits.

CABLES AND OUTLET BOXES INSTALLATION (*cont.*)

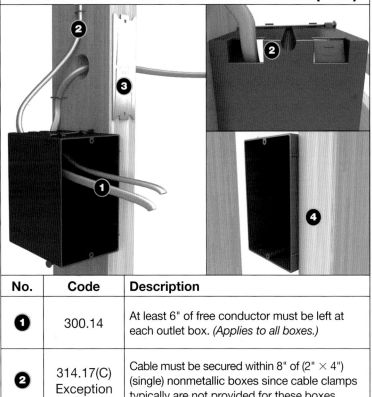

No.	Code	Description
❶	300.14	At least 6" of free conductor must be left at each outlet box. *(Applies to all boxes.)*
❷	314.17(C) Exception	Cable must be secured within 8" of (2" × 4") (single) nonmetallic boxes since cable clamps typically are not provided for these boxes.
❸	300.4	Steel plates must be provided to protect cables routed closer than 1¼" from the edge of wood framing members.
❹	314.20	Boxes should be mounted flush with finished surface. Most boxes feature built-in depth gauges for easy installation.

> **YOU SHOULD KNOW: NEC 314.20 & 314.21**
>
> • If the box is recessed (not flush) in a combustible wall, listed box extenders must be used. Boxes installed in noncombustible walls must not be recessed more than ¼".
>
> • Drywall and plaster around boxes utilizing flush type covers must be repaired so that there are no gaps greater than ⅛" at the edges of the box.

ROUGHING IN RECEPTACLE OUTLETS

HOW IT WORKS

Step	Directions	
1	Remove outer sheathing from NM cable. Leave at least 1/4" of sheathing inside of the outlet box. **314.17(C)** Leave at least 6" of free conductor at each outlet box. **300.14**	
2	Make a V with the grounding conductors and grab with the tip of side cutting pliers.	
3	Twist pliers clockwise, 4–5 times, while slightly pulling on the grounding conductors to make a neat connection.	

If using crimp sleeves follow steps 4 and 5.
If using grounding wire connectors skip to step 6.

ROUGHING IN RECEPTACLE OUTLETS (*cont.*)

HOW IT WORKS

Step	Directions	
❹	If using a crimp sleeve, install sleeve over grounding conductors and crimp with a crimping tool. If only one device is installed, the remaining grounding conductors may be cut off after the crimp sleeve. If multiple devices are installed in the box, leave as many grounding conductors as there are devices.	
❺	With the grounding conductors properly connected, the device is ready to be installed.	
❻	If using a grounding wire connector, cut off any grounding conductors that will not be used before installing the connector. With the grounding conductors properly connected the device is ready to be installed.	

Step 6 is not necessary if using crimp sleeves.

STANDARD MOUNTING HEIGHTS

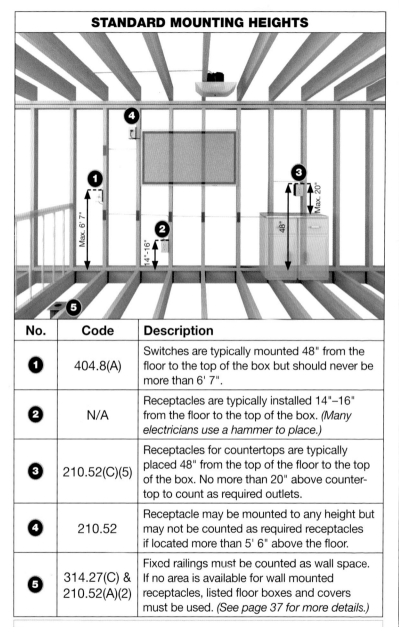

No.	Code	Description
❶	404.8(A)	Switches are typically mounted 48" from the floor to the top of the box but should never be more than 6' 7".
❷	N/A	Receptacles are typically installed 14"–16" from the floor to the top of the box. *(Many electricians use a hammer to place.)*
❸	210.52(C)(5)	Receptacles for countertops are typically placed 48" from the top of the floor to the top of the box. No more than 20" above countertop to count as required outlets.
❹	210.52	Receptacle may be mounted to any height but may not be counted as required receptacles if located more than 5' 6" above the floor.
❺	314.27(C) & 210.52(A)(2)	Fixed railings must be counted as wall space. If no area is available for wall mounted receptacles, listed floor boxes and covers must be used. *(See page 37 for more details.)*

YOU SHOULD KNOW:
- Lights are typically centered over the room or area where lighting is desired.

LOCKOUT PROVISIONS AND GROUNDING

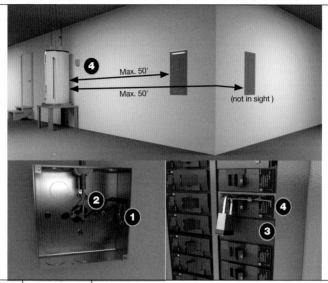

No.	Code	Description
❶	110.12(A)	All unused holes in equipment must be effectively closed.
❷	250.4	All conductive materials, such as metallic outlet boxes, enclosing electrical conductors or equipment must be grounded.
❷	250.8	Solder is not allowed for grounding connections. Listed connectors, screws that engage at least 2 threads, or terminal bars must be used for all grounding connections.
❸	408.7	Unused openings for circuit breakers must be closed.
❹	422.31	Appliances rated over 300 watts or $\frac{1}{8}$ hp without a marked OFF position must have a disconnecting means located within sight of the appliance or the circuit breaker must be capable of being locked in the OPEN position *(cord cap for cord and plug connected appliances may serve as the disconnecting means)*.

▶ **YOU SHOULD KNOW:**
  • In Sight From is defined in Article 100 as visible and no more than 50' apart.

TYPICAL CALCULATIONS

No.	Item	Quantity		Volume		Total
❶	Hots	____	×	____	=	____
❷	Neutrals	____	×	____	=	____
❸	Grounds	____	×	____	=	____
❹	Switch or Receptacle	____ × 2	×	____	=	____
Total Volume Requirements for this Box =						

No.	Code	Description		
❶	314.16(B)(1)	Count the number of "Hot" wires in the box and enter the quantity.		
❷	314.16(B)(1)	Count the number of "Neutral" wires in the box and enter the quantity.		
❸	314.16(B)(5)	Typically enter "1" for the quantity of ground wires. *(Code only requires the largest ground wire to be counted.)* The volume (in cubic inches) will be based on the largest ground wire in the box.		
❹	314.16(B)(4)	Enter the quantity of switches or receptacles. *(Devices count as 2 wires for each gang required for mounting. For devices wider than a standard 2" box, enter the quantity of gangs required for mounting the device.)*		
❺	Table 314.16(B)	For Column two, enter the Volume (in cubic inches) based on the wire size using the following table. *(The volume for switches or receptacles will be based on the largest wire connected to the device.)*	**Wire Size**	**Volume in cu. in.**
			#14	2
			#12	2.25
			#10	2.5
			#8	3
			#6	5

▶ **YOU SHOULD KNOW:**

- If box contains more than one wire size, simply use the Calculation Table above for each wire size but remember only the largest ground must be calculated and each switch or receptacle counts as two wires based on the largest wire connected to the device.

REQUIREMENTS

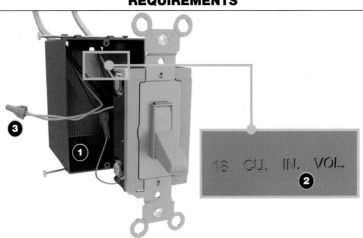

Item	Quantity		Volume	Total
Hots	2	×	2.25	4.5
Neutrals	2	×	2.25	4.5
Grounds	1	×	2.25	2.25
Switch	2	×	2.25	4.5
		Total Volume Needed for this Box =		15.75

EXAMPLE: Calculate the required box size for two 12/2 NM cables *(one hot, one neutral, and one ground in each cable)* terminating in a box with a switch.

No.	Code	Description
1	314.16	Boxes must be of sufficient size to provide free space for all enclosed wires and devices.
2	314.16(A)(2)	Nonmetallic boxes will be marked with their volume.
3	314.16(B)	No volume allowance required for small fittings such as locknuts or bushings.

▶ **YOU SHOULD KNOW: 314.16**
- This is a typical box fill calculation for a nonmetallic outlet box containing small conductors. *(See 314.16 for complete box fill requirements.)*

OVERHEAD

No.	Code	Description
1	590.6	All receptacle outlets must be GFCI protected including 240 volt receptacles.
2	230.23	Service conductors must be sized according to the load but in no case smaller than #8 AWG Copper or #6 AWG Aluminum.
3	250.24(B)	Main bonding jumper must be installed.
4	250.50 & 250.52(A)(5)	8' Ground rod must be driven.
5	230.24	Overhead services must provide a clearance of at least 12' from grade to the lowest point of the conductors when crossing over residential property.
6	230.54(C)	Point of attachment must be below weatherhead.
7	N/A	All temporary services must be properly braced and supported.

> **YOU SHOULD KNOW:**
> * See pages 61–70 for more details.
> * Always check local requirements and location with area inspector and local power supplier.

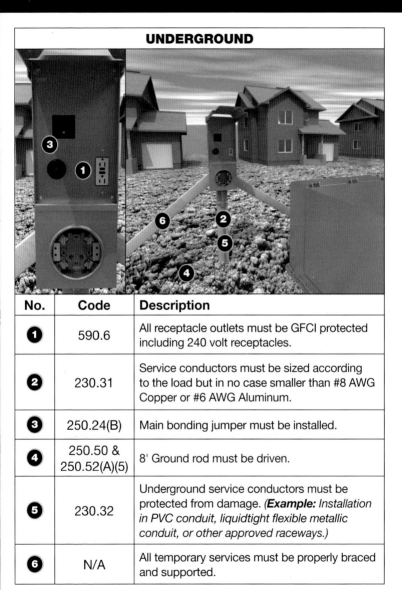

UNDERGROUND

No.	Code	Description
1	590.6	All receptacle outlets must be GFCI protected including 240 volt receptacles.
2	230.31	Service conductors must be sized according to the load but in no case smaller than #8 AWG Copper or #6 AWG Aluminum.
3	250.24(B)	Main bonding jumper must be installed.
4	250.50 & 250.52(A)(5)	8' Ground rod must be driven.
5	230.32	Underground service conductors must be protected from damage. *(**Example:** Installation in PVC conduit, liquidtight flexible metallic conduit, or other approved raceways.)*
6	N/A	All temporary services must be properly braced and supported.

> **YOU SHOULD KNOW:**
> • See pages 61–70 for more details.
> • Always check local requirements and location with area inspector and local power supplier.

RESIDENTIAL AREA REQUIREMENTS

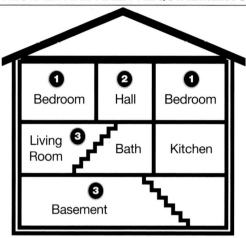

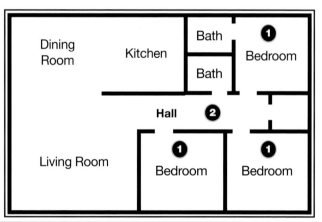

No.	2010 Code	Description
1	2010 NFPA 72 29.5.1.1(1)	Smoke detectors must be located in each sleeping room.
2	2010 NFPA 72 29.5.1.1(2)	Smoke detectors must be located outside of each sleeping room within 21' of any door to a sleeping room.
3	2010 NFPA 72 29.5.1.1(3)	Smoke detectors must be located on each level of the dwelling unit including basements, but NOT including crawl spaces or uninhabitable attics.

WIRING METHOD

No.	Code	Description
❶	210.12(B)	120 volt power supply (2 conductor with ground NM cable) from AFCI protected circuit.
❷	2010 NFPA 72 29.5.2.1.1	Smoke detectors must be interconnected so that actuation of one alarm will activate alarms on all smoke detectors.
❸	2010 NFPA 72 29.6.2 & 29.6.3	Smoke detectors must receive their primary power from the building wiring and be provided with battery backup. Wiring must be permanent and without a disconnecting switch other than a circuit breaker.
❹	N/A	When one detector is actuated, the red conductor carries a signal to all other smoke detectors causing them to activate.

▶ **YOU SHOULD KNOW:**

- Interconnection of smoke detectors is typically made with a 3 conductor with ground NM cable routed to each successive smoke detector.
- Do not install smoke detectors within 3' of the tip of a ceiling fan or an AC register.
- Do not install smoke detectors within 3' of a door leading to a bathroom containing a shower or tub.
- Do not install smoke detectors within 10' of a fixed cooking appliance unless listed for the application.

SMOOTH CEILING

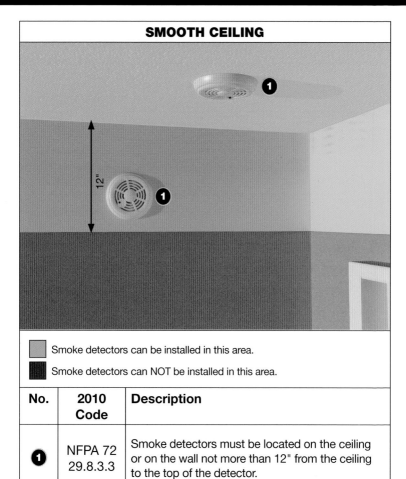

Smoke detectors can be installed in this area.

Smoke detectors can NOT be installed in this area.

No.	2010 Code	Description
1	NFPA 72 29.8.3.3	Smoke detectors must be located on the ceiling or on the wall not more than 12" from the ceiling to the top of the detector.

YOU SHOULD KNOW:

- **110.3(B).** Requires all manufacturers instructions to be followed when installing listed equipment such as smoke detectors.

- NFPA 72 mandates smoke detector placement. Manufacturer's instructions provide these rules for smoke detector placement. Read instructions carefully.

- Previous editions of NFPA 72 and some manufacturer's instructions require that smoke detectors located on the ceiling be placed at least 4" from the sidewall and detectors on the sidewall be located within a zone that extends from 4"–12" below the ceiling. Be sure to check with your local inspector to verify the edition of the NFPA 72 that is enforced in your area. If manufacturer's instructions specify the location of the detectors, those instructions must be followed.

PLACEMENT

PEAKED CEILINGS 2010 NFPA 72 29.8.3.1

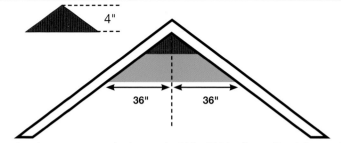

Smoke detectors must be located within 36" horizontally of the peak, but no closer than 4" vertically to the peak.

TRAY CEILINGS 2010 NFPA 72 29.8.3.4(10)

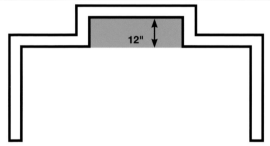

Smoke detectors must be installed on the highest point of the ceiling, or on the sloped portion of the ceiling, not farther than 12" below the highest point of the ceiling.

SLOPED CEILINGS 2010 NFPA 72 29.8.3.2

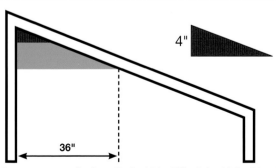

Smoke detectors must be installed within 36" of the high side of the ceiling, but not closer than 4" from the adjoining wall surface.

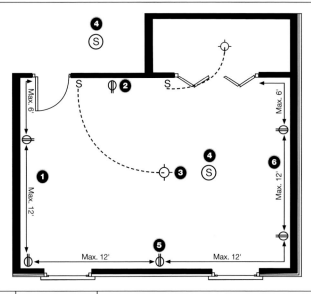

No.	Code	Description
❶	210.52(A)(1)	Receptacle outlets must be spaced so that no point along the wall is farther than 6' from the nearest outlet.
❷	210.52(A)(2)	Wall space 2 feet or greater must have a receptacle outlet.
❸	210.70(A)(1)	Switched lighting outlet is required. Lighting fixture or switched plug is acceptable.
❹	2010 NFPA 72 29.5.1.1(1)	Smoke detectors required in every sleeping room, outside of every sleeping room, and on every level of the home. *(See pages 27–30.)*
❺	210.12(B)	Arc-fault protection required for all outlets. *(Outlets as defined by the NEC includes all lighting fixtures, smoke detectors and receptacles.)*
❻	406.11	All 15 and 20 amp, 120 volt receptacle outlets must be listed tamper resistant type.
2011 NEC	406.12	Tamper resistant receptacles are not required for receptacle outlets that are located more than 5½' above the floor, are part of a listed appliance, or where installed in a dedicated space for an appliance that is not easily moved.

CLOTHES CLOSET

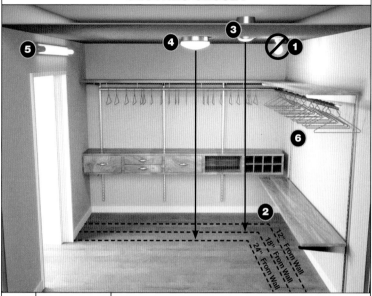

No.	Code	Description
❶	410.16(A)	Open bulb incandescent lighting fixtures are not allowed in clothes closets.
❷	410.2	Storage space is considered 12" (minimum) from side and rear walls or the shelf width, whichever is greater.
❸	410.16(C) (3), & (4)	Recessed incandescent lights with completely enclosed lamps, LED lights with completely enclosed light sources, and recessed fluorescent lights must be 6" from nearest storage space *(18" from side and rear walls)*.
❹	410.16(C)(1)	Surface mounted incandescent or LED lights with completely enclosed light sources must be 12" from nearest storage space *(24" from side and rear walls)*.
❺	410.16(C)(2)	Surface mounted fluorescent lights must be 6" from nearest storage space *(18" from side and rear walls)*.
❻	410.16(C)(5)	Surface mounted fluorescent or LED lights identified for such use may be installed within the storage space.

> ## YOU SHOULD KNOW:

- **240.24(D).** Panelboards not allowed in clothes closets.

- **210.12(B).** Receptacle outlets and lighting fixtures are allowed but NOT required in closets, but if they are installed they must be AFCI protected.

- **410.6.** All lighting fixtures must be listed *(tested by a Nationally Recognized Testing Laboratory such as UL or ETL)*.

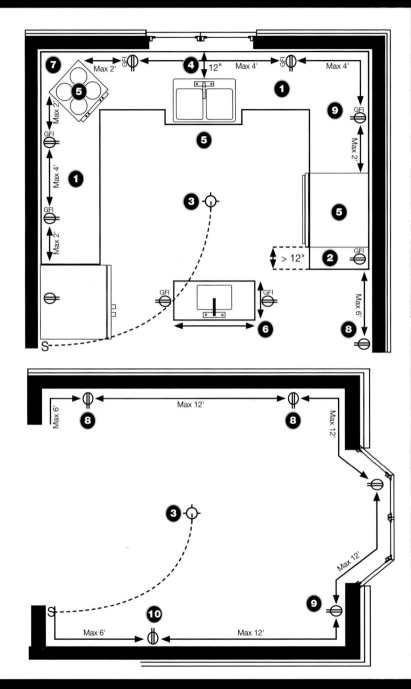

LAYOUT (*cont.*)

No.	Code	Description
1	210.52(C)(1)	Receptacles must be placed so that no point along the countertop is further than 2' from a receptacle.
2	210.52(C)(1)	Counter spaces longer than 1' require a receptacle.
3	210.70(A)(1)	At least one wall switch controlled lighting fixture required.
4	210.52(C)(4)	If spacing behind the sink is less than 12", one plug is required within 2' of each side of the sink.
	210.52(C)(1) Exception	If spacing behind the sink is 12" or greater, this area must be counted as counter space and receptacle must be installed as required in bullet 1 above.
5	210.52(C)(4)	Counter spaces divided by range tops, refrigerators, or sinks must be considered separate spaces.
6	210.52(C)(2)	An island or peninsula space requires a receptacle only if larger than 24" × 12". If a sink or cooktop is installed on the island, and the width of the counter behind the appliance is less than 12" a receptacle must be installed on both sides. *(Locate receptacle no more than 12" below cabinet.)*
7	210.52(C)(4)	If spacing behind cooktop is less than 18", one receptacle is required within 2' of each side of the cooktop.
	210.52(C)(1) Exception	If spacing behind cooktop is 18" or greater, this area must be counted as wall space and receptacles must be installed as required in bullet 1 above.
8	210.52(A)	No point along the wall may be further than 6' from a receptacle, if not counter space.
9	406.11	All 15 and 20 amp, 120 volt receptacle outlets must be listed tamper resistant type.
10	210.12(B)	All 15 and 20 amp, 120 volt outlets in dining rooms and eating areas must be AFCI protected. *(Not required in kitchens.)*

▶ **YOU SHOULD KNOW:**

- **210.8(A)(6).** Receptacles installed to serve countertop surfaces must be GFCI protected.
- **210.11(C)(1).** All receptacles in kitchen area must be on a 20 amp circuit.
- **210.52(B)(3).** Minimum of two 20 amp circuits required to supply kitchen counter top receptacles.
- **210.52(B)(1 & 2).** No other device allowed on the receptacle circuits (*Example: lights or appliances*). Receptacle for refrigerator is allowed on the small appliance circuits or may be supplied from a dedicated 15 or 20 amp circuit.

APPLIANCES – RANGE, RANGEHOOD AND MICROWAVE

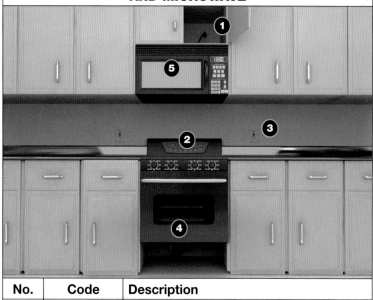

No.	Code	Description
1	422.16(B)(4)	Receptacles for microwaves should be mounted in the cabinet directly above the appliance. GFCI protection is not required.
2	210.19(A)(3)	Ranges rated 8³/₄ kW or more must be supplied by a minimum 40 amp circuit.
3	210.52(C)(5)	Receptacles to be mounted no more than 20" above the counertop.
4	250.142(B)	4 wire circuit required for cord and plug connected free standing ranges. *(Typically wired with 50 amp circuit, #6/3 with ground CU NM cable.)*
5	422.16(B)(4)	Range hoods and built-in microwaves may be cord and plug connected if the receptacle is accessible and supplied from an individual circuit. *(Microwave ovens are usually supplied by a dedicated 20 amp circuit, 12/2 with ground CU NM cable.)*

▶ **YOU SHOULD KNOW:**

• Range receptacles should be located within range space and within 6" to 12" of the floor. This allows the range to fit close to the wall while allowing the receptacle to be easily reached with removal of the bottom drawer.

APPLIANCES – DISHWASHER AND DISPOSAL

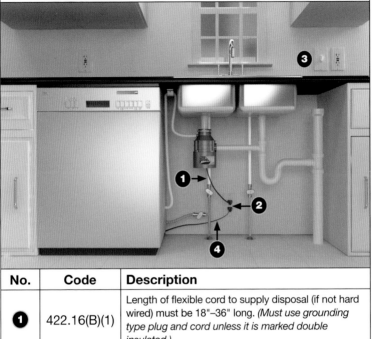

No.	Code	Description
❶	422.16(B)(1)	Length of flexible cord to supply disposal (if not hard wired) must be 18"–36" long. *(Must use grounding type plug and cord unless it is marked double insulated.)*
❷	422.16 (B)(2)(4)	Receptacle must be located in the space occupied by the dishwasher or adjacent cabinet.
❸	422.32, 430.104 & 430.109(C)	Switch is allowed for disposal disconnecting means if not cord and plug connected and the switch is marked with an ON and OFF position.
❹	422.16(B)(2)	Length of flexible cord to supply dishwasher (if not hard wired) must extend 3'–4' from the rear of the appliance. *(Must use grounding type plug and cord unless it is marked double insulated.)*

▶ YOU SHOULD KNOW:

- **240.5(B)(1) and Table 400.5(A).** Minimum 18 AWG flexible cord may be used for Maximum 10 amp load as long as appliance is listed for this cord size. Use larger cord size for higher wattage appliances.

- **422.31(B).** If dishwasher is hard wired and not within sight of the circuit breaker, a lockout bracket must be installed on the circuit breaker. *(See page 22.)*

LAYOUT

Max. 6'

Max. 6'

Max. 6'

Max. 12'

Max. 6'

Max. 12'

6 Floor Mounted
Receptacle

Max. 6'

S

Max. 6'

GFI

No.	Code	Description
1	210.52(A)(1)	Receptacle outlets must be spaced so that no point along the wall is farther than 6' from the nearest outlet. A fireplace, doorway or similar opening is not required to be counted as wall space (*receptacles must be located within 6' of each side of these spaces*).
2	210.52(A)(2)	Wall space 2' or greater must have a receptacle outlet.
3	210.52(A)(2)	Fixed panels such as the portion of a sliding glass door that is not movable must be considered wall space.
4	210.70(A)(1)	Switched lighting outlet is required. Lighting fixture or switched plug is acceptable.
5	210.8(A)(7)	GFCI protection required for all 120 volt 15 or 20 amp receptacles located within 6' of a wet bar sink.
6	210.52(A)(3) 314.27(C)	If wall space is not available, receptacles must be installed on the floor. Receptacles installed on the floor must be within 18" of the wall to be counted as a required receptacle (*must use listed floor boxes and covers*).

WETBAR AND TRACK LIGHTING

No.	Code	Description
1	210.8(A)(7)	GFCI protection required for all 120 volt, 15 or 20 amp receptacles located within 6' of a wet bar sink. *(No exceptions for single outlet or dedicated appliances.)*
2	406.11	All 15 and 20 amp, 120 volt receptacle outlets must be listed tamper resistant type.
	210.2(B)	All 15 and 20 amp, 120 volt outlets in living rooms and bonus rooms must be AFCI protected.
3	210.52(A)(2)	Space occupied by counter top must be counted as wall space and satisfy the requirement for placing receptacles along the wall *(no point along the wall can be further than 6' from a receptacle).*
4	410.151(C)(8)	Track lighting is not allowed less than 5' above the floor unless protected from damage.
2011 NEC	210.8	All GFCI devices must be installed in a readily accessible location. *(Must be able to reach GFCI device quickly without having to use a ladder or climb over obstacles.)*
2011 NEC	210.52(A)(2)	Wall space occupied by fixed cabinets does not have to be counted as wall space when installing receptacle outlets. *(See bullet 1 on page 37.)*

▶ **YOU SHOULD KNOW: NEC**

- **410.151.** Track lighting must be hard wired. Cord connection is not allowed.
- **410.154.** Track lighting must be supported at a minimum of two points every 4.

No.	Code	Description
1	210.70(A)(2)	Wall switch controlled lighting outlet required.
2	210.70(A)(2)	Where there are 6 or more risers on the stairs, a wall switch to control the lighting outlet is required at each level. *(3 way or 4 way switching required.)*
3	210.52(H)	Hallways 10' or more in length must have at least 1 receptacle outlet.
	406.11	All 15 and 20 amp, 120 volt receptacle outlets must be listed tamper resistant type.
	210.12(B)	All 15 and 20 amp, 120 volt outlets in hallways and stairwells must be AFCI protected.
4	2010 NFPA 72 29.5.1.1(2)	Smoke detectors must be located outside of each sleeping room within 21' of any door to a sleeping room. *(See pages 27–30.)*
2011 NEC	210.52(I)	Foyers that are not part of a hallway and have an area greater than 60 square feet, must have a receptacle outlet installed in each wall space 3' or wider.

▶ **YOU SHOULD KNOW:**

- **210.52(H).** Length of odd shaped hallways, such as L or T shaped, must be measured along the center of the hallway to a point passing through a doorway.

LAUNDRY ROOM

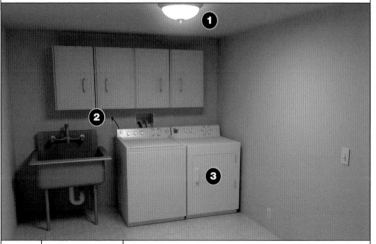

No.	Code	Description
➊	210.70(A)(1)	Switched lighting outlet required.
➋	210.11(C)(2)	Dedicated 20 amp circuit required for laundry receptacle outlets. No other devices allowed on the laundry receptacle circuit. (**Example:** *lights or receptacles located in another room.*)
	210.52(F)	At least 1 receptacle outlet (20 amp circuit) required in every dwelling unit for laundry purposes.
	210.8(A)(7)	GFCI protection required for all 120 volt, 15 and 20 amp receptacles located within 6' of a utility or laundry sink. *(No exceptions for single outlet or dedicated appliances.)*
	406.11	All 15 and 20 amp, 120 volt receptacle outlets must be listed tamper resistant type.
➌	250.142(B)	4 wire 30 amp circuit required for electric dryers.

▶ **YOU SHOULD KNOW:**

- New ranges and clothes dryers are required by UL standards to come from the factory with a bonding strap joining the ground and neutral terminals. This strap must be removed to comply with the code. Follow manufacturer's instructions to make cord connections to new ranges and clothes dryers.

- AFCI protection is not required for outlets installed in laundry rooms.

TYPICAL BATHROOM

No.	Code	Description
1	210.70(A)(1)	At least one wall switch controlled lighting outlet required.
2	210.8(A)(1) & 210.52(D)	All receptacles must be GFCI protected (on a 20 amp circuit) and at least one must be installed within 3' of each sink.
	210.11(C)(3)	No other devices allowed on the receptacle circuit, unless the dedicated 20 amp circuit supplies only one bathroom. *(Example: lights, vent fans, etc.)*
	210.11(C)(3)	Different bathroom receptacles (only) may be served from one 20 amp circuit.
3	410.10(D)	No part of a lighting track, hanging lighting fixture or ceiling fan allowed within a 3' horizontal by 8' vertical zone above the threshhold of a shower or the rim of a bathtub. *(Recessed or surface mounted lighting fixtures are allowed in this zone, but must be marked for use in a damp location. If subject to shower spray, they must be marked for use in a wet location.)*
	404.4 & 406.8(C)	Receptacles or switches not allowed within or directly over a tub or shower space.

HYDROMASSAGE BATHTUB

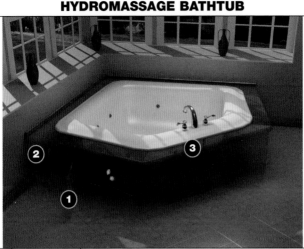

No.	Code	Description
①	680.73	Access is required to all electrical equipment for hydromassage bathtubs.
②	680.71	Hydromassage bathtub electrical equipment must be on an individual branch circuit and be protected by a readily accessible GFCI device.
	406.11	All 15 and 20 amp, 120 volt receptacle outlets must be listed tamper resistant type.
③	680.74	Metal piping systems and all grounded metal parts in contact with the water must be bonded together with a solid copper wire not smaller than #8. *(Pump motors usually have an exterior lug for making this connection.)*

▶ **YOU SHOULD KNOW:**

- **110.3(B).** Vent fans installed directly above showers must be GFCI protected *(manufacturer's instructions)*.

- **Article 100.** The NEC defines readily accessible as capable of being reached quickly for operation, renewal or inspections without requiring those to whom ready access is requisite to climb over or remove obstacles or to resort to portable ladders, and so forth.

GARAGE

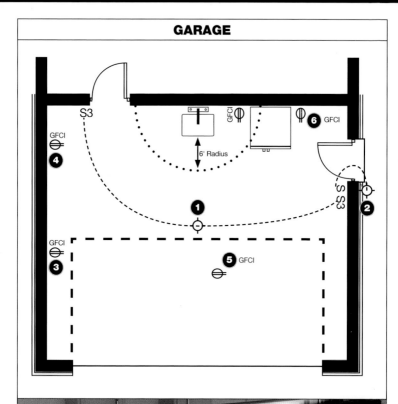

6' Radius

S3

GFCI **4**

GFCI **3**

1

GFCI **5** GFCI

S3

2

GFCI **6** GFCI

GARAGE (cont.)

No.	Code	Description
1	210.70(A)(2)	Wall switch controlled lighting outlet is required.
2	210.70(A)(2)	Switched lighting outlet required on exterior to illuminate all personnel entrances. *(Not required for vehicle doors, such as overhead garage doors.)*
3	210.52(G)	At least 1 receptacle outlet *(in addition to those for specific equipment such as garage door openers or washing machines)* is required in all attached garages.
4	210.8(A)(2)	GFCI protection is required for all 15 and 20 amp 120 volt receptacles.
5	406.11	All 15 and 20 amp, 120 volt receptacle outlets must be listed tamper resistant type.
6	210.8(A)(7)	ALL 15 and 20 amp, 120 volt receptacles must be GFCI protected including outlets for refrigerators, garage door openers and washing machines.

▶ **YOU SHOULD KNOW:**

- Some jurisdictions require smoke detectors and/or carbon monoxide detectors for garages. Check your local municipality requirements.

- **210.52(G)**. Requirements for detached garages with electric power are the same as those for attached garages.

UNFINISHED BASEMENT

No.	Code	Description
1	210.70(A)(3)	Wall switch controlled lighting outlet is required.
	210.70(A)(2)	Switched lighting outlet required on exterior to illuminate all personnel entrances. *(Not required for vehicle doors, such as overhead garage doors.)*
2	210.52(G)	At least one receptacle outlet is required in unfinished basements.
3	210.52(G)	Where a portion of the basement is finished into a habitable room, one receptacle outlet is required in each unfinished portion.
4	210.8(A)(5)	GFCI protection is required for all 15 and 20 amp, 120 volt receptacles.
5	406.11	All 15 and 20 amp, 120 volt receptacle outlets must be listed tamper resistant type.

▶ **YOU SHOULD KNOW:**

  • Basements that are finished into living space must comply with the appropriate requirements for the specific type of room involved.

No.	Code	Description
6	210.8(A)(5) Exception	GFCI protection not required for a receptacle supplying a permanently installed burglar or fire alarm system.
7	210.8(A)(5)	ALL 120 volt, 15 and 20 amp receptacles must be GFCI protected including outlets for refrigerators, garage door openers, and washing machines unless they are supplying a permanently installed burglar or fire alarm system.
8	334.15	NM cable must closely follow the surface of the building or be protected from damage. *(Must protect horizontal runs of exposed cable.)*
9	334.15(C)	NM cable used on the wall of an unfinished basement is permitted to be installed in listed conduit. *(Must use insulating bushing at the point where the cable enters, secure the cable within 12" of the conduit, and the sheathing of the cable must extend not less than ¼" into the outlet box.)*

▶ YOU SHOULD KNOW: 334.15(C)

- Cables containing wires smaller than two #6 or three #8 are not allowed to be secured directly to ceiling joists. *(Must be routed through bored holes or installed on running boards.)*

STORAGE AREA

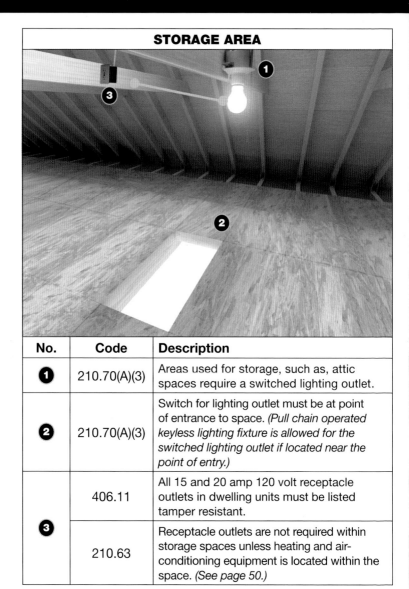

No.	Code	Description
①	210.70(A)(3)	Areas used for storage, such as, attic spaces require a switched lighting outlet.
②	210.70(A)(3)	Switch for lighting outlet must be at point of entrance to space. *(Pull chain operated keyless lighting fixture is allowed for the switched lighting outlet if located near the point of entry.)*
③	406.11	All 15 and 20 amp 120 volt receptacle outlets in dwelling units must be listed tamper resistant.
	210.63	Receptacle outlets are not required within storage spaces unless heating and air-conditioning equipment is located within the space. *(See page 50.)*

YOU SHOULD KNOW:

- **2010 NFPA 72 29.8.3.4(2).** Smoke detectors should NOT be installed in uninhabitable attics.

OUTSIDE LIGHTING AND RECEPTACLES

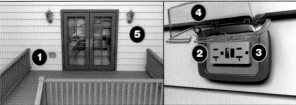

No.	Code	Description
1	210.52(E)	Receptacle outlet required at the front and rear outside of every home. *(No higher than 6' 6" above grade.)*
	210.52(E)(3)	Receptacle outlet required within the perimeter of each balcony, deck or porch with a usable area of 20 square feet or greater that is accessible from the inside of the home. *(No higher than 6' 6" above the floor surface.)*
2	210.8(A)(3)	All 15 and 20 amp, 120 volt receptacles must be GFCI protected.
	406.11	All 15 and 20 amp, 120 volt receptacle outlets must be listed tamper resistant type.
	406.8(A)	All 15 and 20 amp, 120 and 240 volt nonlocking receptacles in wet or damp locations must be listed weather resistant type.
3	406.8(A)	Receptacles installed outdoors must have a weatherproof cover *(flip covers acceptable for damp locations).*
4	406.8(B)(1)	Outdoor receptacles in wet locations, unprotected from rainfall must have a cover that is weatherproof whether or not a cord is plugged in *("in-use" or "bubble" covers).*
5	210.70(A)(2)	Switched lighting outlet required on exterior to illuminate all personnel entrances *(not required for vehicle doors, such as overhead garage doors).*
2011 NEC	210.52(E)(3)	ALL balconies, decks and porches that are accessible from the inside of the home must have a receptacle outlet installed within their perimeter. *(Exception for small areas has been removed.)*

CRAWL SPACE

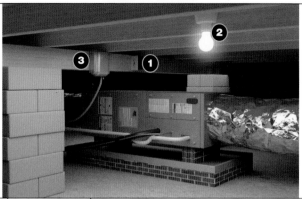

No.	Code	Description
❶	210.63	Receptacle required if heating or air conditioning equipment is present. Receptacle must be located on the same level and within 25' of the equipment.
	210.8(A)(4)	Receptacles must be GFCI protected.
❷	210.70(A)(3)	Lighting outlet required if space is used for storage or contains equipment that may need servicing. Light must be located at or near the equipment (*Example: heating or air conditioning equipment*).
	210.70(A)(3)	Switch for lighting outlet must be at point of entrance to space.
❸	440.14	Disconnecting means for HVAC equipment must be located within sight and readily accessible from the equipment. (*May NOT be mounted on panels designed to allow access to the equipment or where the nameplate will be obscured.*)

> **YOU SHOULD KNOW:**

- Cables containing wires smaller than two #6 or three #8 are not allowed to be secured directly to ceiling joists. (*Must be routed through bored holes or installed on running boards.*)

- **Article 100.** The NEC defines a "damp location" as a location protected from weather and not subject to saturation with water or other liquids but subject to moderate degrees of moisture. Some inspectors and jurisdictions consider crawl spaces a damp location and will require weatherproof enclosures for electrical equipment installed in these areas.

INSIDE EQUIPMENT

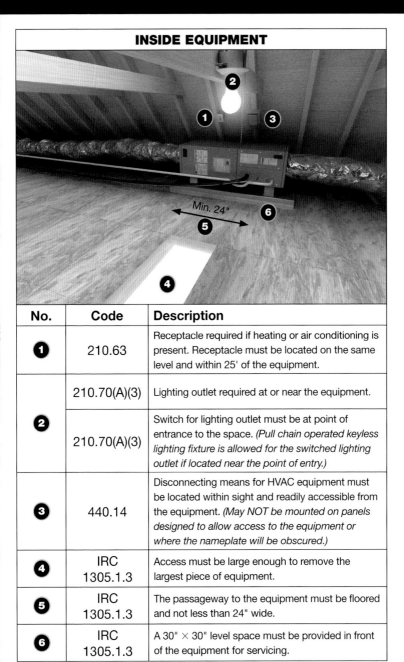

No.	Code	Description
1	210.63	Receptacle required if heating or air conditioning is present. Receptacle must be located on the same level and within 25' of the equipment.
2	210.70(A)(3)	Lighting outlet required at or near the equipment.
	210.70(A)(3)	Switch for lighting outlet must be at point of entrance to the space. *(Pull chain operated keyless lighting fixture is allowed for the switched lighting outlet if located near the point of entry.)*
3	440.14	Disconnecting means for HVAC equipment must be located within sight and readily accessible from the equipment. *(May NOT be mounted on panels designed to allow access to the equipment or where the nameplate will be obscured.)*
4	IRC 1305.1.3	Access must be large enough to remove the largest piece of equipment.
5	IRC 1305.1.3	The passageway to the equipment must be floored and not less than 24" wide.
6	IRC 1305.1.3	A 30" × 30" level space must be provided in front of the equipment for servicing.

OUTSIDE EQUIPMENT

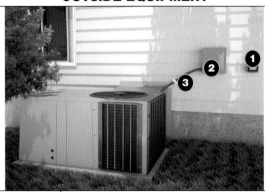

No.	Code	Description
❶	210.63	Must have receptacle on the same level and located within 25' of all HVAC equipment.
	210.8(A)(3)	Receptacle must be GFCI protected.
	406.8(A)	All 15 and 20 amp, 120 and 240 volt nonlocking receptacles in wet or damp locations must be listed weather resistant type.
	406.11	All 15 and 20 amp, 120 volt receptacle outlets must be listed tamper resistant type.
❷	440.14	Disconnecting means for HVAC equipment must be located within sight and readily accessible from the equipment. *(May NOT be mounted on panels designed to allow access to the equipment or where the nameplate will be obscured.)*
	110.26	Must maintain clear working space about all electrical equipment that may require servicing while energized. *(Do NOT mount disconnect directly behind equipment.) (See page 57.)*
❸	300.6	Conduit installed outside in a wet location must be listed for use outdoors. *(Liquidtight flexible metal conduit or liquidtight flexible nonmetallic conduit is most commonly used for this application.)*
	300.9 & 310.8(C)	Conductors installed outdoors must be listed for use in wet locations. *(NM cable not allowed for outdoor use even if installed in conduit; conductors with a W in wire type designator [**Example:** THWN wire] must be used.)*

No.	Code	Description
4	350.6 & 356.6	All fittings used must be listed for use with each specific conduit type. *(Liquidtight nonmetallic fittings required for use with liquidtight nonmetallic conduit.)*
5	350.60 & 356.60	Equipment grounding conductors must be installed when liquidtight flexible metallic or nonmetallic conduit is used and flexibility after installation is required.
	408.40	Bonding screw or jumper must be installed to connect equipment grounding bar to metallic enclosure.
	250.119	Grounding conductors must have green insulation, green insulation with yellow stripes, or be bare. *(Only grounding conductors larger than #6 may be reidentified by taping or marking at termination points.)*

YOU SHOULD KNOW:

- A light is not required at the HVAC equipment if installed outside.

- **430.113.** If motor operated equipment is supplied from more than one source *(**Example:** two disconnects serving one piece of equipment)*, a disconnect must be provided for each source and a permanent warning sign must be provided on or adjacent to each disconnect.

CONDUCTOR SIZING

Table 1: Sizing "Hot" Conductors for HVAC Equipment (based on Table 310.16)

Wire (Conductor) Size	Allowable Ampacity (in Amps)	
	Copper	Aluminum*
#14	20	—
#12	25	20
#10	35	30
#8	50	40
#6	65	50
#4	85	65
#3	100	75
#2	115	90
#1	130	100
1/0	150	120
2/0	175	135
3/0	200	155
4/0	230	150
250 kcmil	255	205

*Most HVAC equipment specifies the use of copper supply conductors only.

Table assumes all terminal provisions and conductors are listed for 75°C.

See page 16 for sizing NM or SE cables.

Table 2: Sizing Grounding Conductors (based on Table 250.122)

Circuit Breaker or Fuse Size (in amps)	Ground Wire Size	
	Copper	Aluminum*
15	#14	#12
20	#12	#10
30–60	#10	#8
70–100	#8	#6
110–200	#6	#4
225–300	#4	#2

*Most HVAC equipment specifies the use of copper supply conductors only.

250.122(A). Grounding conductors are never required to be larger than the "Hot" conductors.

AIR HANDLERS

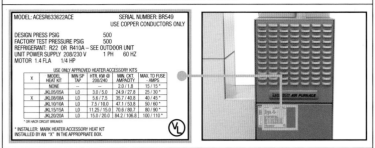

	MODEL HEAT KIT	MIN SP TAP	HTR. KW @ 208/240	MIN. CKT. AMPACITY	MAX. TD FUSE - AMPS
	NONE	--		2.0 / 1.8	15 / 15 *
	JKL05/05A	LO	3.0 / 5.0	24.9 / 27.8	25 / 30 *
X	JKL08/08A	LO	5.6 / 7.5	35.7 / 40.8	40 / 45 *
	JKL10/10A	LO	7.5 / 10.0	47.1 / 53.8	50 / 60 *
	JKL15/15A	LO	11.25 / 15.0	70.6 / 80.7	80 / 90 *
	JKL20/20A	LO	15.0 / 20.0	84.2 / 106.8	100 / 110 *

MODEL: ACESR633622ACE   SERIAL NUMBER: BR549
USE COPPER CONDUCTORS ONLY

DESIGN PRESS PSIG   500
FACTORY TEST PRESSURE PSIG   500
REFRIGERANT: R22 OR R410A – SEE OUTDOOR UNIT
UNIT POWER SUPPLY 208/230 V   1 PH   60 HZ
MOTOR 1.4 FLA   1/4 HP

USE ONLY APPROVED HEATER ACCESSORY KITS

* OR HACR CIRCUIT BREAKER

* INSTALLER: MARK HEATER ACCESSORY HEAT KIT
INSTALLED BY AN "X" IN THE APPROPRIATE BOX.

Step	Directions
1	Determine minimum circuit ampacity from equipment nameplate. Be sure to get data from appropriate voltage column. Most units are suitable for use with 208 or 240 volt systems. Each column will have a different value. *(In this example, the minimum circuit ampacity for this unit is 40.8 amps when operating at 240 volts.)*
2	Determine conductor size from Table 1 *(see page 53)* based on the minimum circuit ampacity found in step 1. Most HVAC equipment is listed for use with copper conductors only. Aluminum conductors are rarely allowed to be used. *(In this example, with a minimum circuit ampacity of 40.8 amps we must use #8 copper supply conductors.)*
3	Determine circuit breaker or fuse sizing from equipment nameplate. Be sure to get data from appropriate voltage column. *(In this example, the maximum fuse or circuit breaker size is 45 amps.)*
4	Determine grounding conductor size from Table 2 (see page 53) based on the circuit breaker or fuse size. Grounding conductors are never required to be larger than the "Hot" conductors. *(In this example, we used a 45 amp circuit breaker so a #10 copper grounding conductor is required.)*

▶ YOU SHOULD KNOW:

- Most air handlers or furnaces installed indoors are generally supplied with NM or SEU cable and grounding conductors will be appropriately sized based on cable size. *(See page 16 to properly size NM or SE cable.)*

- Gas furnaces are typically supplied by a single 15 or 20 amp, 120 volt circuit. A general use switch may be used as the disconnecting means if located within sight of the equipment and marked with an ON and OFF position.

- Field installed electric heat kits must be identified on equipment nameplates. This will be accomplished by marking appropriate heat kit installed on equipment name plate or installing factory supplied label over existing data.

CONDENSING UNITS

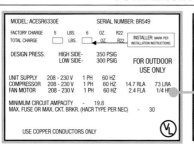

MODEL: ACESR6330E   SERIAL NUMBER: BR549

| FACTORY CHARGE | 5 | LBS. | 6 | OZ. | R22 | |
| TOTAL CHARGE | | LBS. | | OZ. | R22 | INSTALLER: MARK PER INSTALLATION INSTRUCTIONS |

DESIGN PRESS.   HIGH SIDE-   350 PSIG
                  LOW SIDE-   300 PSIG   FOR OUTDOOR
                                             USE ONLY

UNIT SUPPLY   208 - 230 V   1 PH   60 HZ
COMPRESSOR   208 - 230 V   1 PH   60 HZ   14.7 RLA   73 LRA
FAN MOTOR   208 - 230 V   1 PH   60 HZ   2.4 FLA   1/4 HP

MINIMUM CIRCUIT AMPACITY   -   19.8
MAX. FUSE OR MAX. CKT. BRKR. (HACR TYPE PER NEC)   -   30

USE COPPER CONDUCTORS ONLY

Step	Directions
1	Determine minimum circuit ampacity from equipment nameplate. Be sure to get data from appropriate voltage column if marked. *(In this example, the minimum circuit ampacity for this unit is 19.8 amps.)*
2	Determine conductor size from Table 1 *(see page 53)* based on the minimum circuit ampacity found in step 1. Most HVAC equipment is listed for use with copper conductors only. Aluminum conductors are rarely allowed to be used. *(In this example, with a minimum circuit ampacity of 19.8 amps we must use #14 copper supply conductors.)*
3	Determine circuit breaker or fuse sizing from equipment nameplate. *(In this example, the maximum fuse or circuit breaker size is 30 amps.)* **Note 240.4(G) & 440.22(C):** Motor loads and motor operated HVAC equipment are not subject to the same rules as non-motor operated equipment. Overcurrent protection will be sized based on nameplate data not necessarily on conductor size. *(Overcurrent protection may never exceed the value marked by the manufacturer on the equipment nameplate.)*
4	Determine grounding conductor size from Table 2 *(see page 53)* based on the circuit breaker or fuse size. Grounding conductors are never required to be larger than the "Hot" conductors. *(In this example we used a 30 amp circuit breaker and the Table on page 53 requires the use of a #10 copper grounding conductor but 250.122(A) allows the grounding conductor to be no larger than the "Hot" conductors. In this case, the "Hot" conductors are #14 so the grounding conductor is not required to be larger than #14 copper.)*

> **YOU SHOULD KNOW:**
> * Motors are rated 115 or 230 volts but are suitable for use with 120 or 240 volt systems.

PACKAGE UNITS

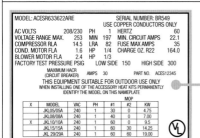

MODEL: ACESR633622ARE			SERIAL NUMBER: BR549			
			USE COPPER CONDUCTORS ONLY			
AC VOLTS	208/230	PH	1	HERTZ		60
VOLTAGE RANGE MAX.	253	MIN	197	MIN. CIRCUIT AMPS		22.1
COMPRESSOR RLA	14.5	LRA	82	FUSE MAX AMPS		35
COND. MOTOR FLA	1.6	HP	1/4	CHARGE OZ. R22		164.0
BLOWER MOTOR FLA	2.4	HP	1/3			
FACTORY TEST PRESSURE PSIG		LOW SIDE	150	HIGH SIDE	300	

MAXIMUM HACR (CIRCUIT BREAKER) AMPS 30   PART NO. ACES12345

THIS EQUIPMENT SUITABLE FOR OUTDOOR USE ONLY
WHEN INSTALLING ONE OF THE ACCESSORY HEAT KITS PERMANENTLY
IDENTIFY THE MODEL ON THIS NAMEPLATE

X	MODEL	VAC	PH	#1	#2	KW
	JKL05/05A	240	1	30	0	4.75
	JKL08/08A	240	1	40	0	7.00
X	JKL10/10A	240	1	60	0	9.5
	JKL15/15A	240	1	60	30	14.25
	JKL 29/29A	240	1	60	60	19.00

MOP at top spanning #1 #2

MOP = MAXIMUM OVERCURRENT PROTECTION

Step	Directions
1	Determine minimum circuit ampacity from equipment nameplate. Be sure to get data from appropriate voltage column. Most units are suitable for use with 208 or 240 volt systems. Each column will have a different value. *(In this example, the minimum circuit ampacity for this unit is 79.3 amps when operating at 240 volts.)*
2	Determine conductor size from Table 1 *(see page 53)* based on the minimum circuit ampacity found in step 1. Most HVAC equipment is listed for use with copper conductors only. Aluminum conductors are rarely allowed to be used. *(In this example, with a minimum circuit ampacity of 79.3 amps we must use #4 copper supply conductors.)*
3	Determine circuit breaker or fuse sizing from equipment nameplate. Be sure to get data from appropriate voltage column. *(In this example, the maximum fuse or circuit breaker size is 90 amps.)*
4	Determine grounding conductor size from Table 2 *(see page 53)* based on the circuit breaker or fuse size. Grounding conductors are never required to be larger than the "Hot" conductors. *(In this example, we used a 90 amp circuit breaker so a #8 copper grounding conductor is required.)*

▶ **YOU SHOULD KNOW:**

- **250.122(C).** Some package units equipped with electrical heating may be supplied by multiple circuits. Each circuit must be sized according to the steps above but only one grounding conductor is required if all circuits are installed in the same conduit. When only one grounding conductor is used, its size is based on the largest circuit feeding the equipment.

- Field installed electric heat kits must be identified on equipment nameplates. This will be accomplished by marking appropriate heat kit installed on equipment nameplate or installing factory supplied label over existing data.

WORK SPACE

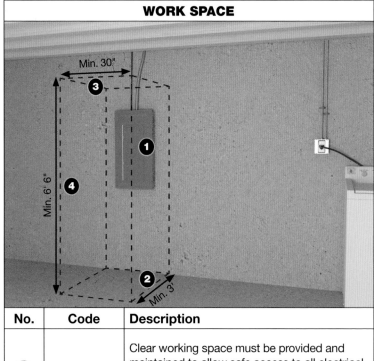

No.	Code	Description
1	110.26	Clear working space must be provided and maintained to allow safe access to all electrical equipment that may require examination or servicing while energized.
2	110.26(A)(1)	A minimum of 3' deep clear space must be provided and maintained in front of equipment operating at 120/240 volts.
3	110.26(A)(2)	A clear space of at least 30" wide (or the width of the equipment – whichever is greater) must be maintained. *(Equipment doors must be able to open at least 90 degrees.)*
4	110.26(E)	At least 6' 6" of clear headroom must be provided and maintained.

▶ YOU SHOULD KNOW:

- This requirement applies to all electrical equipment, including panelboards, disconnects, and similar equipment. This allows an operator or technician to service equipment safely and escape if electrocution or arc flash occurs.

CONDUCTOR SIZING		
Sizing Conductors for Feeders and Services (based on Table 310.15(B)(6))		
Service or Feeder Rating (amps)	**Conductor (AWG or kcmil)**	
	Aluminum or Copper-Clad Aluminum	**Copper**
100	#2	#4
110	#1	#3
125	1/0	#2
150	2/0	#1
175	3/0	1/0
200	4/0	2/0
225	250	3/0
250	300	4/0
300	350	250
350	500	350
400	600	400

Code	Description
215.2(A)(1), 230.23(A), & 230.31(A)	Service and feeder conductors must be sized according to the calculated load. *(See pages 75–76 for load calculation example.)*
310.15(B)(6)	Service and feeder conductors for single phase 120/240 volt dwelling unit electrical systems may be sized per Table 310.15(B)(6). *(See Table above.)*
310.15(B)(6)	Table 310.15(B)(6) may ONLY be used for dwelling unit services or feeders if the service or feeder supplies ALL loads that are associated with the dwelling unit. This table may NOT be used to size conductors supplying power panelboards such as HVAC disconnects or subpanels supplying only a portion of the load associated with the home. *(See page 16 for sizing NM or SE cable for these loads or Table 1 on page 53 when using individual conductors in a raceway.)*

SUBPANELS

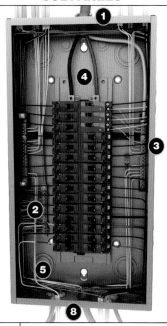

No.	Code	Description
1	250.142	Subpanels must be supplied with 4 conductors *(isolate grounds and neutrals)*.
	215.2(A)(1)	Size feeder conductors to subpanels that supply lighting and appliance loads per the calculated load and according to Table on page 58.
2	408.40	Equipment grounding bar must be bonded to the subpanel enclosure.
	110.3(B)	Multiple grounding conductors may be installed under the same terminal if they are the same size and maximum number of conductors does not exceed manufacturer's instructions.
3	408.41	Only one neutral conductor allowed under each terminal.
	408.40	Bonding jumper to neutral bar must not be installed.
4	408.36	Main breaker not required for subpanels if overcurrent protection is provided for feeder conductors.

SUBPANELS (*cont.*)

No.	Code	Description
5	200.7(C)	White insulated conductors used as "Hots" must be identified at all termination points.
6	110.3(B)	Correct breaker types must be used for each panelboard. *(GE breakers cannot be installed in Siemens Panels.)*
	210.12(B)	All 15 and 20 amp, 120 volt circuits supplying dining rooms, living rooms, bedrooms, sunrooms, closets, hallways, or similar areas must be AFCI protected.
7	408.4	Each circuit must be clearly identified as to it's purpose. *(No 2 circuits may be labeled the same.)*
	408.4	When labeling circuits, no circuit shall be identified in a manner that may be subject to changes in occupancy *(**Example:** "Keith's Bedroom").*
8	312.5(C)	Each cable must be secured to the panelboard enclosure. *(Use listed cable connectors.)*

▶ **YOU SHOULD KNOW:**

• Subpanels are frequently located throughout the home to shorten the length of branch circuit cables.

• SER cable is most commonly used to supply subpanels.

• An example load calculation for subpanels may be found on pages 75–76. Typical subpanels supply limited loads such as lighting and small appliances and are commonly supplied by 100 amp circuits using #2 copper SER cable or 1/0 aluminum SER cable.

SERVICE EQUIPMENT

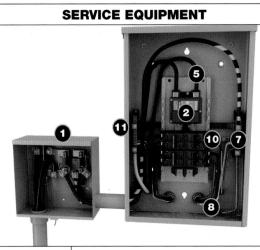

No.	Code	Description
1	230.2	Only one service allowed per building.
	230.79(C)	Minimum 100 amp service required for a dwelling unit.
	230.71(A)(1)	Disconnecting means for service must be located either outside or inside the building nearest the point of entrance of the service conductors.
2	230.71(A)	No more than 6 circuit breakers may be used to disconnect all service conductors. *(Main breakers are typically used and commonly required by local authorities.)*
3	230.70(B)	Service equipment must be clearly marked to identify it as a service disconnect.
4	408.4	Each circuit must be clearly identified as to its purpose. *(No two circuits may be labeled the same.)*
	408.4	When labeling circuits, no circuit shall be identified in a manner that may be subject to changes in occupancy *(**Example:** "Keith's HVAC Unit").*

▶ **YOU SHOULD KNOW:**

- Most common service size for single family dwellings is 200 amps. However, it is not uncommon to find 400, 600 or even 800 amp services on larger homes.

- Always verify local requirements with area inspector and utility company.

- Utility company will specify location of service equipment and metering.

SERVICE EQUIPMENT (*cont.*)

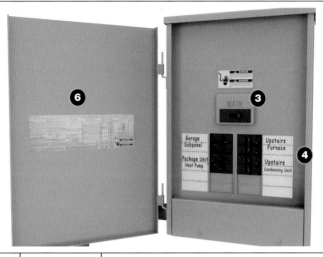

No.	Code	Description
5	310.15(B)(6)	Size service conductors for 120/240 volt single phase dwelling unit services per Table on page 58.
6	230.70(C)	Each service disconnect must be suitable for the conditions in which it is installed. (***Example:*** *Use NEMA type 3R panel in wet locations.)*
7	250.24(B)	Main bonding jumper must be installed. *(Grounds and neutral bars tied together and connected to enclosure.)*
8	312.5(C)	Each cable must be secured to the panelboard enclosure. *(Use listed cable connectors.)*
9	200.6	Neutral conductor must be identified by white insulation or white markings at termination points.
10	200.7(C)	White insulated conductors used as "Hots" must be identified at all termination points.
11	408.41	Only one neutral conductor allowed under each terminal.

> **YOU SHOULD KNOW:**
>
> - **230.70(A)(2) & 240.4(E).** Service equipment or other panelboards are not allowed in residential bathrooms or clothes closets.

SERVICE EQUIPMENT (*cont.*)

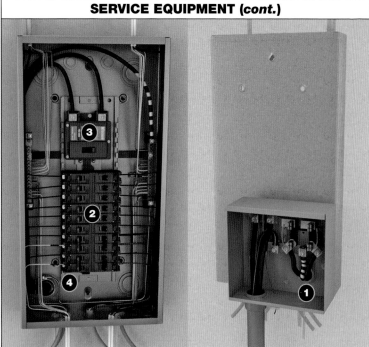

No.	Code	Description
❶	300.4(G)	Plastic bushings must be used at connectors to protect conductors from abrasion.
❷	110.3(B)	Must use correct breaker types for each panelboard. (Cannot install GE breakers in Siemens panels.)
❸	230.70(A)	Disconnecting means for service must be located either outside or inside the building nearest the point of entrance of the service conductors.
❹	250.92(B)	Bonding bushings and jumpers are required to be used for metal conduit entering panel through concentric or eccentric knockouts.

▶ **YOU SHOULD KNOW:**

• See previous page for additional requirements.

• Clear workspace must be maintained around all panelboards. (See page 57.)

UNDERGROUND RACEWAYS

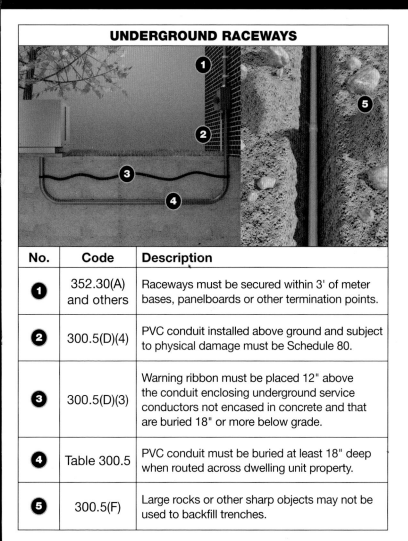

No.	Code	Description
1	352.30(A) and others	Raceways must be secured within 3' of meter bases, panelboards or other termination points.
2	300.5(D)(4)	PVC conduit installed above ground and subject to physical damage must be Schedule 80.
3	300.5(D)(3)	Warning ribbon must be placed 12" above the conduit enclosing underground service conductors not encased in concrete and that are buried 18" or more below grade.
4	Table 300.5	PVC conduit must be buried at least 18" deep when routed across dwelling unit property.
5	300.5(F)	Large rocks or other sharp objects may not be used to backfill trenches.

> **YOU SHOULD KNOW:**
>
> • Typically a 2" minimum conduit size is required for 200 amp services. Local power companies specify size and type of conduit that must be used. Some power companies supply conductors from transformer to meter base but most require a pulling string installed in conduit.

OVERHEAD SERVICES

No.	Code	Description
1	230.54(C)	Point of attachment for service drop must be below the weatherhead.
	230.24(A) Exception #4	Vertical clearances from the edge of roof are not required to be maintained for final conductor span where the service drop is attached to the side of a building.
2	230.9(A)	Must maintain at least 3' clearance from open service conductors to openable windows, porches, balconies or similar locations.
3	230.28	Service masts used as support for service drop conductors must be adequately braced or supported by guy wires. *(Local rules usually apply.)*
	230.28	Only power cables are allowed to be attached to service mast. *(No cable or phone drops allowed.)*

OVERHEAD SERVICES (*cont.*)

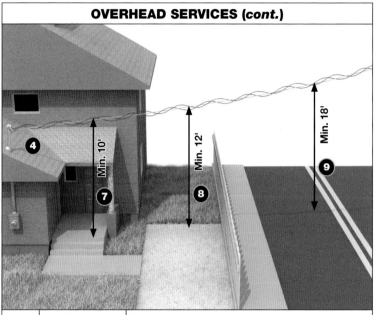

No.	Code	Description
❹	230.24(A) Exception #3	Service conductors are generally required to be 8' above the roof but if the service conductors pass over only the overhang portion of the roof for a horizontal distance of 4' or less and terminate at a "through the roof" mast assembly, the clearance may be reduced to 18".
❺	N/A	Height of meter base is determined by local power company but is generally located 5–6' above grade.
❻	344.30 and others	Conduit must be supported within 3' of each termination and every 10'. *(Conduit extending through a roof should be secured within 3' of the point where conduit passes through the roof.)*
❼	230.24(B)(1)	Service conductors must be at least 10' above sidewalks.
❽	230.24(B)(2)	Service conductors must be at least 12' above residential driveways or yards.
❾	230.24(B)(4)	Service conductors must be at least 18' above public streets.

GROUNDING ELECTRODE CONDUCTORS

Grounding at Services
(based on Table 250.66)

Size of Service Conductors		Size of Grounding Electrode Conductors	
CU	**AL**	**CU**	**AL**
#2 or smaller	1/0 or smaller	#8	#6
#1–1/0	2/0–3/0	#6	#4
2/0–3/0	4/0–250	#4	#2
4/0–350	300–500	#2	1/0
400–600	600–900	1/0	3/0
700–1100	1000–1750	2/0	4/0
Over 1100	Over 1750	3/0	250

Code	Description
250.24(D)	All grounding electrode conductors must be connected to the service equipment enclosure. *(Main breaker panel.)*
250.66	Grounding electrode conductors must be sized per Table above with the exception of ground rods, rebar in footer, or ground rings.
250.64(A)	Where used outside, aluminum grounding conductors may not be terminated within 18" of the earth.
250.64(C)	Grounding electrode conductors may not be spliced with split bolts. *(Only irreversible splices such as crimped connections may be used.)*
250.102(C)	Bonding jumpers used for bonding around concentric or eccentric knockouts must be sized based on the table above.
250.52(B)	Metal underground gas piping or aluminum electrodes are not permitted as grounding electrodes.
250.94	An intersystem bonding termination for connecting grounding conductors of other systems *(cable TV and telephone)* must be installed external to enclosures at the service disconnecting means.

YOU SHOULD KNOW:

• A properly installed grounding system is essential to a safe electrical system. Some of the purposes for grounding include limiting voltages caused by lightning, stabilizing the voltage under normal operating conditions and ensuring the facilitation of overcurrent devices under ground-fault conditions.

REBAR IN FOOTER

No.	Code	Description
❶	250.50	All of the following grounding electrodes that are present at each building must be bonded together to form the grounding electrode system.
	250.52(A)(1)	Metal underground water piping. *(See page 69.)*
	250.52(A)(5)	Ground rods or pipes. *(See page 70.)*
	250.52(A)(3)	Rebar in footer.
❷	250.66(B)	Grounding electrode conductor that is the sole connection to rebar in the footer is not required to be larger than #4 copper.
❸	250.68(A) Exception #1	Connection to rebar in footer is not required to be accessible if encased in concrete.
	250.70	Connection to rebar in footer must be made by means of a listed connector. *(If encased in concrete, connector must be listed for concrete encasement and direct burial.)*

METALLIC WATER PIPING

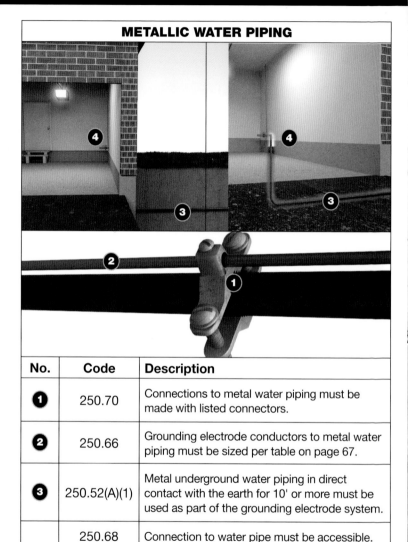

No.	Code	Description
1	250.70	Connections to metal water piping must be made with listed connectors.
2	250.66	Grounding electrode conductors to metal water piping must be sized per table on page 67.
3	250.52(A)(1)	Metal underground water piping in direct contact with the earth for 10' or more must be used as part of the grounding electrode system.
4	250.68	Connection to water pipe must be accessible.
	250.52(A)(1)	Connection to metal water piping must be made within the first 5' upon entering the building.

> **YOU SHOULD KNOW:**

- Refer to page 67 for general rules regarding connections of grounding electrode conductors.

GROUND RODS

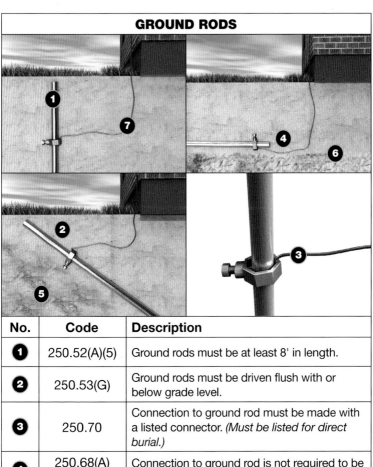

No.	Code	Description
1	250.52(A)(5)	Ground rods must be at least 8' in length.
2	250.53(G)	Ground rods must be driven flush with or below grade level.
3	250.70	Connection to ground rod must be made with a listed connector. *(Must be listed for direct burial.)*
4	250.68(A) Exception #1	Connection to ground rod is not required to be accessible.
5	250.53(G)	Where rock bottom is encountered, the rod may be driven at an angle not to exceed 45° from vertical.
6	250.53(G)	Where rock bottom is encountered at an angle up to 45°, the rod may be buried in a trench at least 30" deep.
7	250.66(A)	Portion of grounding electrode conductor that is the sole connection to the rod is not required to be larger than #6 copper. *(Check with local power company. Many power companies require larger conductors.)*

DETACHED BUILDINGS

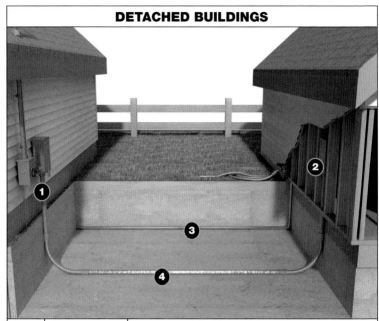

No.	Code	Description
❶	300.5(D)(4)	PVC conduit installed above ground and exposed to damage must be Schedule 80.
❷	210.52(G) 210.70(A)(2)	Receptacle and lighting requirements are the same for attached and detached garages. *(See page 44.)*
❸	250.94	An intersystem bonding termination for connecting grounding conductors of other systems *(cable TV and telephone)* must be installed external to enclosures at the disconnecting means.
❸	250.32(A)	Grounding electrode system requirements are the same as for service equipment. *(See pages 67–70.)*
	250.32(A) Exception	Where the detached garage is served by only one branch circuit a grounding electrode system is not required. *(1 – 20 amp, 120 volt circuit installed for lighting and general use receptacles would not require a ground rod or connection to metal piping.)*

DETACHED BUILDINGS (*cont.*)

Minimum Burial Depths for Residential Property

	Direct Buried Cables	Rigid Metal or IMC Conduit	PVC or Other Raceways	120 Volt 20 Amp GFCI Circuits	30 Volts or Less
All Locations Not Specified	24"	6"	18"	12"	6"
Below 2" of Concrete	18"	6"	12"	6"	6"
Below 4" of Concrete (not driveways)	18"	4"	4"	6" (direct burial) 4" (in raceway)	6" (direct burial) 4" (in raceway)
Under Driveways	18"	18"	18"	12"	18"

No.	Code	Description
	300.5(A)	Raceways and direct buried conductors must be installed to minimum burial depths provided in the table above.
4	300.12	Conduit systems must be continuous between cabinets or boxes. (*Breaks in conduit are not allowed.*)
	300.10	Metal conduit systems must be electrical continuous between metal enclosures. (*Cannot have metal conduit between 2 sections of nonmetallic conduit.*)
	300.5(B)	Conductors installed underground must be listed for use in a wet location. (*Conduit in the ground is considered a wet location. Must use conductors with a W in the wire type designator.* **Example:** *THWN or XHHW.*)
5	300.5(F)	Large rocks or other sharp objects may not be used to backfill ditches.

PANELS

No.	Code	Description
❶	225.31 225.32 225.33 225.36	Must have main breaker panel, or not more than 6 circuit breakers, installed either outside the building or inside nearest the point of entrance to the building. Panelboard must be listed as suitable for use as service equipment. *(Service equipment rated disconnect is not required for a single branch circuit. A general use snap switch may be used as the disconnect. Requirements are the same for service equipment. See pages 61–66.)*
❷	225.5 225.39	Feeder conductors and overcurrent protection to detached garages must be based on the calculated load. *(See pages 75–76 for load calculation example.)*
❸	250.32(B)	120/240 volt single phase feeders to new detached buildings must contain an equipment grounding conductor *(4 wire feeder).*

PANELS (*cont.*)

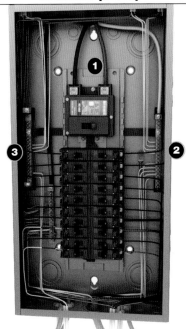

No.	Code	Description
❶	225.5	Conductors supplying detached buildings must be sized from Table 1 on page 53.
❷	250.32(B)	Isolate grounds and neutrals in disconnect and connect any grounding electrode conductors to the equipment grounding bar. (*Do NOT install main bonding jumper.*)
❸	250.32(B)	Equipment grounding conductors must be sized based on Table 2 on page 53.
2011 NEC	210.52(G)	Requirements for accessory buildings with electric power are the same as those for attached and detached garages. (*See pages 43–44.*)

LOAD CALCULATIONS

Step	Directions

LIGHTING AND GENERAL USE RECEPTACLES: 220.82(B)(1)

❶ Square footage ___**2,570**___ × 3 = ___**7,710**___ va

(Do NOT count the square footage of open porches, garages or unused or unfinished spaces not adaptable for future use.)

SMALL APPLIANCE AND LAUNDRY CIRCUITS: 220.82(B)(2)

❷ Number of circuits ___**4**___ × 1,500 = ___**6,000**___ va

(Each home must have a minimum of 2 – 20 amp small appliance circuits serving the kitchen and dining area receptacles and at least 1 – 20 amp circuit to supply the laundry receptacles. The total minimum number of circuits required for these receptacles will always be 3. If more circuits are provided, the number of circuits provided must be used.)

APPLIANCE AND MOTOR LOADS: 220.82(B)(3) AND (4)

❸ Enter the nameplate rating of all appliances and motor loads in each box. If appliance is not listed, use the miscellaneous row. Do NOT enter heating or cooling loads into this area.

Cooktop	**N/A** va
Oven	**N/A** va
Range	**10,500** va
Water Heater	**4,500** va
Dishwasher	**950** va
Garbage Disposal	**550** va
Clothes Dryer	**4,800** va
Trash Compactor	**N/A** va
Microwave Oven	**1,580** va
Garage Door Opener	**450** va
Ice Maker	**N/A** va
Hydromassage Tub Motor	**875** va
Miscellaneous	**N/A** va
TOTAL	**37,915** va

LOAD CALCULATIONS (*cont.*)

Step	Directions

TOTAL STEPS 1 – 3: 220.82(B)

④

1. Total of Loads __37,915__ – 10,000 va = __27,915__
2. Line 1 __27,915__ × 40% = __11,166__
3. Line 2 __11,166__ + 10,000 va = __21,166__

Total of Line 3 __21,166__ va

(*This is the total calculated load for all appliances, lighting and receptacles. If the calculation is for a feeder supplying a lighting and appliance subpanel and no heating or air conditioning equipment is fed from this panel the load calculation is complete. Proceed to step 6 for sizing overcurrent protection and conductors.*)

HEATING AND AIR CONDITIONING LOADS: 220.82(C)

⑤

Enter the nameplate rating of all applicable heating or air conditioning equipment. Most homes will only have 1 or 2 of the selections.

A. Air Conditioning Equipment __N/A__ va
B. Heat Pump without supplemental Heating __N/A__ va
C. Supplemental heating for heat pump
  __17,500__ × 65 % = __11,375__
  + Heat pump compressor __9,120__ = __20,495__ va
D. Electric Space Heating (*< 4 separately controlled units*)
  __N/A__ × 65 % = __N/A__ va
E. Electric Space Heating (*4 or more separately controlled units*)
  __N/A__ × 40 % = __N/A__ va
F. Electric thermal storage heating __N/A__ va

CALCULATE TOTAL SERVICE OR FEEDER LOAD: 220.82(A)

⑥

Total of Line 3 from Step 4 __21,166__ va
Enter only the largest load from Step 5 + __20,495__ va

Total Calculated Service or Feeder Load = __41,661__ va

CALCULATE SERVICE OR FEEDER SIZE

⑦

Total Calculated Load __41,661__ va / 240 volts = __173.6__ amps

This calculation resulted in a calculated load of 173 amps, which would require a minimum service size of 175 amps and the use of 1/0 copper or 3/0 aluminum service conductors; however, future expansion and additions should be considered and a 200 amp service with 2/0 copper or 4/0 aluminum service conductors would be recommended.

OHM'S LAW

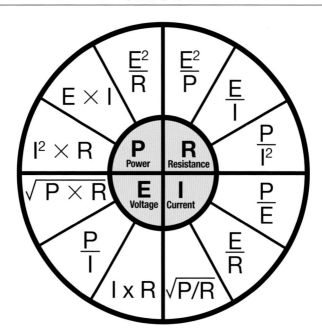

P = Power – Measured in Watts
I = Intensity of Current – Measured in Amps
E = Electromotive Force (EMF) – Measured in Volts
R = Resistance – Measured in Ohms

Examples:
A hair dryer is rated at 1,800 watts and connected to a 120 volt circuit. What is the current draw of this hair dryer?
I = P / E = 1,800 / 120 = 15 amps

What is the resistance of the hair dryer?
R = E² / P = (120 × 120) / 1,800 = 14,400 / 1,800 = 8 ohms

For 3 phase circuits the same formulas may be used but the system voltage must be increased by the square root of 3. (E × 1.732)

Example:
A 3 phase water heater is rated at 3,200 watts and connected to a 208 volt circuit. What is the current draw of this circuit?
I = P / (E × 1.732) = 3,200 / (208 × 1.732) = 3,200 / 360 = 8.9 amps